ISBN-13: 978-1533168078
ISBN-10: 1533168075

A Book-Souvenir on

INDUSTRIAL GLOBALIZATION and
ENVIRONMENTAL AWARENESS

Proceedings of the **NATIONAL SEMINAR** held by
Department of Chemistry,
Government PG College, Ambala Cantt - Haryana - India
on **11th February 2016.**

Edited By

Dr AVTAR SINGH RAHI
M.Sc. NET M.Phil. Ph.D. M.Ed. Ph.D. MBA PGDHRM PGDJ

Seminar Convener and Head
Department of Chemistry, Government PG College,
Ambala Cantt. (India)

Chemistry of Industrial Globalization, Environmental Pollution and its Chem-Biological Significance

National Seminar sponsored by
Directorate of Higher Education, Haryana

Citation:

Name of the Author(s), **2016,** Topic of the Presentation, *Industrial Globalization and Environmental Awareness (ISBN-13: 978-1533168078, Ed. Dr Rahi A.S.)*, *Proceedings of the NATIONAL SEMINAR held by Department of Chemistry, Government PG College, Ambala Cantt - Haryana - India.*

AIR

Chemistry of Industrial Globalization, Environmental Pollution and its Chem-Biological Significance

Proceedings of the **NATIONAL SEMINAR** held by

Department of Chemistry, Government PG College, Ambala Cantt - Haryana - India on 11th February 2016

WATER

EARTH

For those
Who Love Life,
Care for Nature,
Conserve Ecosystem and
Maintain Environmental Composition.
Also for Those
Who Thinks Good,
Spreads Better and
Try to do The Best.

GOVERNMENT POST-GRADUATE COLLEGE, AMBALA CANTT.

(NAAC accredited 'A' Grade Institution)

extends its cordial invitation on the occasion of

NATIONAL SEMINAR

Sponsored by: Directorate of Higher Education, Haryana

on 11th February, 2016 at 10.00 A.M.

Topic: *"Chemistry of Industrial Globalization, Environmental Pollution and Its Chem-Biological Significance"*

Chief Guest

Sh. O.P. Singh
IPS

**Commissioner of Police,
Ambala-Panchkula**

Valedictory Address

Prof. N.S. Atri

Dept. of Botany
Punjabi University, Patiala

Key Note Speaker

Prof. Ravi Shankar

Head, Dept. of Chemistry
Indian Institute of Technology, New Delhi

Resource Person

Dr Inderjeet Singh Sandhu

Professor and Dean,
Chitkara University, Punjab

Dr Avtar Singh Rahi

Convener

Ph:0171-2644503

(M): 9417129908

Email: rahiavtaar@gmail.com

Dr Kamlesh

Principal

GOVERNMENT P.G. COLLEGE, AMBALA CANTT.

Haryana, India

Organizing

One Day

National Seminar

Sponsored by:
Director, Higher Education, Haryana

on

11.02.2016

Chemistry of Industrial Globalization

Environmental Pollution and its

Clean-Biological Significance

ORGANIZING COMMITTEE

Patron   **Dr Subodh Kumar**
Offg Principal

Seminar Convener

Dr AVTAR SINGH RAHI
M.Sc. NET M.Phil. Ph.D. M.Ed. Ph.D
PGDIIRM PGDMBA
Head, Dept. of Chemistry
9417129908/ 0946-403114
rahiavtaar@gmail.com
avtaarrah@rediffmail.com

Seminar Facilitator

Dr Rohini Singh
Head, Dept. of Botany

Organizing Committee Members

Sh. Satish Garg, Head, Dept. of Physics
Sh. Harish Kumar, Dept. of Chemistry
Ms Seema, Dept. of Chemistry
Ms Sandeep Kaur, Dept. of Chemistry
Ms Monika, Dept. of Physics
Ms Komal, Dept. of Botany
Ms Daya, Dept. of Zoology
Sh. Vikas Kumar, Dept. of Zoology
Ms Anju, Dept. of Physics
Ms Sushma, Dept. of Physics
Ms Priyanka, Dept. of Physics

Science Faculty
GOVERNMENT PG COLLEGE, AMBALA CANTT
(NAAC Accredited 'A' GRADE)
Contact No. 0171- 2644503
Fax : 0171-2640853
Website: www.gcambalacantt.com

About the College

Government Postgraduate College, Ambala Cantt was established in 1997. It has a rare distinction of being one of the very few Government Colleges to be accredited 'A' status by NAAC. The institution is a prestigious and high rated one in the academic circles and it stands committed to creating and providing opportunity for the overall development of students that can transform society also. The College has always been pioneering in introducing such new programmes and courses that cater to the growing demands of students in the current competitive scenario. The College imparts education in the disciplines of Humanities, Commerce and Sciences with a vision of National Building. The College has postgraduate degree course in Commerce, Economics, Geography, History, English, Punjabi, Hindi and postgraduate diploma in Mass Communication. The college is running various job oriented courses such as BTM, B.A. (Mass Communication), BCA, B.Sc. and B.Com. (Computer Application) for the purpose of enriching the students with fast changing trends and needs of the Global scenario. With extensive infrastructure and faculty members of the high caliber, the college emphasizes on quality education coupled with value added programmes, co-curricular and extra curricular activities. Department of Commerce is the major strength of the college. The department plays a pivotal role in developing the student community in the field of Commerce, Management and Entrepreneurial competencies.

About the City

Ambala is situated on the North-Eastern edge of Haryana. Ambala is known as City of Scientific Instruments on international level. The city has its historical importance. It is well connected by rail and road network. It is gateway to Chandigarh, Punjab, Himachal Pradesh and Jammu & Kashmir. The city is 52 km from City beautiful Chandigarh, 148 km from Shimla, 198 km from New Delhi and 250 km from Amritsar.

Theme of the National Seminar

Chemistry of Industrialization for Environmental Pollution and its Control

For sustainable development to be achieved, links between the environment and development must be examined. It is also important to consider the end point of development: human well-being. People depend on natural resources for their basic needs, such as food, energy, water and housing. The ability to meet material needs is strongly linked to the provisioning, regulating and supporting services of ecosystems. Global demand for energy and materials keep growing, placing an ever-increasing burden on natural resources, research and the environment. Chemists from all over the world are using their creative and innovative skills to develop new processes, synthetic methods, analytical tools, reaction conditions, catalysts, etc. As the feedstock industry for modern society, the chemical industry plays a major role in the sustainability effort – to advance the science and technology to support the design, creation, processing, use, and disposal of chemical substances that provide a foundation for sustainability. Awareness, creativity, and looking ahead are needed to bring reactions and chemical processes to maximum efficiency. Main goal should be to find ways to produce technology in ways that do not damage or deplete the Earth's natural resources. Chemistry will have much more to offer by becoming more meaningful to humanity, increasing in attractiveness as a career choice, growing to be more worthy of support, spawning new, large economic developments, and progressing to be more interesting and compelling if chemists work to define and follow their natural and unique role in achieving a virtuous civilization that sees broad validity within the community of living things for the claim to continuity of existence in an environment of natural genesis

Sub-Themes

1. Significance of Research in Chemistry to Environment and Community,
2. Chemical and Biological Decomposition of Environmental Pollutants,
3. Role of Research, Industry and Technology on water and Air,
4. Use of Technology in creating Comforts and Luxuries,
5. Role of Needs to explore different applications of Technology,
6. Environmental Contamination and Pollution,
7. Effects of Chemical and other types of Industry on Health,
8. Role of Globalization in Industrial Development,
9. Biological, Physical and Environmental Impacts of Industrialization,
10. Recycling: Use and Impact,
11. Politicization of Globalization and Environmental Impacts,
12. Natural Resources and their Depleting Effects,
13. Global Warming, Green Investment, Green Research, Green Industry, Green Education, And any other Sub-Theme Related to Theme and Topic . . . .

PROGRAMME SCHEDULE

Registration and Tea	0900 A.M. – 0930 A.M.
Inaugural Session	0930 A.M. – 1130 A.M.
Tea Break	1130 A.M. – 1145 A.M.
Technical Sessions	1145 A.M. – 0230 P.M.
Lunch	0230 P.M. – 0300 P.M.
Valedictory Session	0300 P.M. – 0430 P.M.
Tea	0430 P.M.

SEMINAR FEE

Teachers/ Industry Persons: Rs.150/-
Ph.D./M.Phil./M.Sc. Students: Rs.100/-
No Fee for Teacher Organizers of Host College.
Boarding/Lodgings are to be borne by

Participation in the National Seminar

1. Teachers, Industry Persons and Researchers in the field of Chemistry, Botany, Zoology, Physics, Biochemistry, Geophysics, Microbiology, Industrial Sciences, Environmental Sciences, Engineering, Pharmaceuticals, Vaidik Sciences, Geography, Political Sciences, Psychology, etc. can participate in this National Seminar.
2. **Participation, Abstract and Research Paper can be emailed up to 20.01.2016 at rahiavtaar@gmail.com**
3. Participation information should include Name, Designation, Affiliation, Qualifications, Date of Birth, Title of the Paper, Name of Co-Author (if any), Mobile, Email, etc.
4. Each Participant will receive a confirmation of participation in the Seminar by 31.01.2016.
5. **Selected Papers will be published in ISBN online Book.**
6. Any participant can be author of maximum two papers
7. Any paper may have any number of authors but for Seminar Participation and Certificate, only first two authors will be considered.
8. Certificate will be awarded to physically present participants only.
9. **Chairpersons for Technical Sessions will be selected from Delegates whose papers will be received before last date.**
10. Papers can be typed in MS-Word in 4-6 pages, 1.5 space, Times New Roman, Title (size-18), Name-Affiliation (size 14,11), Headings (size 12), Text (size-11), covering Introduction, Literature Search, Materials and Methods, Discussion, Conclusion, Significance and Bibliography.
11. A hard copy of the paper is to be submitted at the time of Registration along-with Copyright Certificate for publication in Book.
12. Please feel free to contact Seminar Convener for any help.

From the desk of Seminar Convener

Economic globalization has numerous implications for environmental sustainability. There are neither theoretical reasons nor adequate or conclusive empirical evidences to show that the relationship between globalization and sustainability is unidirectional or unidimensional. Growing and cumulative scale of human activities has produced environmental effects of a global nature that are not reflected in the markets but that affect global common interests transcending national perspectives. The environmental effects of productive and technological restructuring may be direct or indirect. The environmental implications of globalization are different from the economic ones, in both time and space. The environmental consequences are generally longer-term, with dynamic, cumulative characteristics that are difficult to measure because they are associated in some cases with qualitative parameters. Some examples of such implications are cross-border pollution, effects on global goods, effects on landscape and the loss of scenic beauty, the extinction of species and the loss of biodiversity. Direct environmental effects are generated by the use of new technologies for agriculture industry and energy, by the exploitation of hitherto untapped renewable and non-renewable natural resources, by the creation and dispersion of new biological forms and by the release of new substances into the environment. Indirect environmental effects are generated by the social, economic, political and demographic adjustments driven by the wave of new technology. In the absence of social and ethical regulations, these developments may lead to overexploitation and degradation of regional ecosystems. The implications of these trends for sustainable development need to be assessed. While it may not always be the fault of humans, humans still need to recognize the extent to which they rely on the resources that the natural world provides.

Toxic pollution affects more than 200 million people worldwide, according to *Pure Earth*, a non-profit environmental organization. Pollution is the process of making land, water, air or other parts of the environment dirty and unsafe or unsuitable to use. This can be done through the introduction of a contaminant into a natural environment. Things as simple as light, sound and temperature can be considered pollutants when introduced artificially into an environment. In some of the world's worst polluted places, babies are born with birth defects, children have lost 30 to 40 IQ points, and life expectancy may be as low as 45 years because of cancers and other diseases. Commercial or industrial waste is a significant portion of waste. Industries generate hazardous waste from mining, petroleum refining, pesticide manufacturing and other chemical production. Households generate hazardous waste as well, including paints and solvents, motor oil, fluorescent lights, aerosol cans, and ammunition. Greater production and use of fertilizers and pesticides is causing environmental problems. The EPA states that the most common contaminants in water are bacteria, mercury, phosphorus and nitrogen. These come from the most common sources of contaminants, which include agricultural runoff, air deposition, water diversions and channelization of streams. The artificial warming of water is called thermal pollution. It can happen when a factory or power plant that is using water to cool its operations ends up discharging hot water. This makes the water hold less oxygen, which can kill fish and wildlife. A common type of air pollution happens when people release particles into the air from burning fuels. This pollution looks like soot, containing millions of tiny particles, floating in the air. Another common type of air pollution is dangerous gases, such as sulfur dioxide, carbon monoxide, nitrogen oxides and chemical vapors. These can take part in further chemical reactions once they are in the atmosphere, creating acid rain and smog. Air pollution can take the form of greenhouse gases, such as carbon dioxide or sulfur dioxide, which are warming the planet through the greenhouse effect. People are exposed to a variety of chemicals on a daily basis. Chemicals are used in a variety of locations visited on a regular basis including schools, grocery stores, and dry cleaners. The ways humans cause pollution are limitless.

Economic reforms have had the effect of modernizing the region's agriculture, Industrialization and urbanization, making it more intensive, with the positive result of reducing pressure on natural areas but economic development, particularly in developing areas, is putting pressure on natural resources. Globalization is the process by which all people and communities come to experience an increasingly common economic, social and cultural environment. It affects the balance of economic, political and cultural power between nations, communities and individuals. Prevention is the best measure for controlling chemical pollution. The pollution control boards actively work with industries to help monitor gas emissions and disposal of hazardous chemicals. There are measures taken to ensure that many of the chemicals and gases used in everyday processes are disposed of and recycled correctly. Individuals are encouraged to reduce their Carbon Footprint, essentially the amount of pollution they contribute to the environment, by recycling, choosing environmentally-friendly modes of transportation such as walking and biking, and living an overall green and environmentally conscious lifestyles.

Seminar Convener is thankful to the directorate of higher education, Haryana for the funds provided to organize this seminar. I am thankful to the Principal, Colleagues and Students of my college for the help and support provided. I also acknowledge the support of media friends for covering this knowledgeable event. At the last but not least I am thankful to the Delegates who made this event a grand success.

Dr Avtar Singh Rahi

Details of Delegates participated in the National Seminar entitled "Chemistry of Industrial Globalization, Environmental Pollution and Its Chem-Biological significance"

SrNo	Name (Dr/Sh./Ms)	Affiliation
1	Anil Jindal	R.K.S.D. College, Kaithal
2	Surender Singh	Government PG College, Jind
3	Deepak Sharma	R.G. Government College, Saha
4	Latika	GuruNanak Khalsa College, Karnal
5	Shashi Sharma	GuruNanak Khalsa College, Karnal
6	Ravi Barwal	Government College, Ambala Cantt.
7	Deepika Sherawat	Government College, Ambala Cantt.
8	Pallavi Rani	Government College, Ambala Cantt.
9	Kirti Rani	Government College, Ambala Cantt.
10	Anita Amani	Government College for Women, Rohtak
11	Ashima	A.S.S.M. College, Mukandpur (Punjab)
12	Narinder Anchal	Government College for Girls, Sector-42, Chandgarh
13	Rajender Swain	Government College for Girls, Sector-42, Chandgarh
14	Prerna Goyal	M.M. PG College, Fatehabad
15	Nikhil Khurana	M.M. PG College, Fatehabad
16	Sanjiv Garg	J.M. Institute of Technology, Radhaur
17	Vinod Kumar	M.M. University, Mullana
18	Girish Kumar	M.M. University, Mullana
19	Ramesh Sandhu	C.R. College of Education, Hisar
20	Alka	Government College, Barwala
21	Ravi Juneja	Government College, Ambala Cantt.
22	Rashmi Dhawan	S.A. Jain College, Ambala City
23	Sandeep Kumar	Dyal Singh College, Karnal
24	Sushil Kumar	S.D. College, Ambala Cantt.
25	Anshul Bansal	S.A. Jain College, Ambala City
26	Parisha	Government College, Bhiwani
27	Reena	Government College for Women, Bhiwani
28	Nisha Saini	Government College for Girls, Sector-11, Chandgarh
29	Veenu Saini	Punjab Technical University, Jalandhar
30	Jaipal Deshwal	K.M. Government College, Narwana
31	Neha	R.G. Mahavidyalya, Jind
32	Ajay Kumar	Institute of Pharmaceutical Sciences, Kurukshetra University, Kurukshetra
33	Shilpy Aggarwal	R.K.S.D. College, Kaithal
34	Anil Kumar	Government college, Bawanikhera
35	Anil Kumar	Arya PG College, Panipat
36	Praveen Kumar Yadav	Government College, Kosli (Rewari)
37	Deepika Sherawat	Government College, Ambala Cantt.

38	Pallavi Rani	Government College, Ambala Cantt.
39	Ranjeet Singh	R.G. Government College, Saha
40	Joginder	S.D. College, Ambala Cantt.
41	Pooja Sharma	S.D. College, Ambala Cantt.
42	Ruchi Gupta	Government College, Ambala Cantt
43	Shruti Sharma	Government College, Ambala Cantt.
44	Vikas Sharma	Government College, Ambala Cantt.
45	Manoj Kumar	Government College of Education, Bhiwani
46	Gulshan Singh	S.D. College, Ambala Cantt.
47	Tanuja	O.P.J.S. University, Rajasthan
48	Maninder Kaur	Punjabi University, Patiala
49	Sunita	S.A. Jain College, ambala City
50	S.P. Singh	J.V. PG College, Baraut (Uttar Pradesh)
51	Namrata Jain	Government College, Ambala Cantt.
52	Shakti Kumar	Government College, Ambala Cantt.
53	Savita Rani	Government College, Ambala Cantt.
54	Parmil Kumar	Government College, Ambala Cantt.
55	Gayatri	Government College, Ambala Cantt.
56	Yogita Maheshwari	D.A.V. College, Yamuna Nagar
57	Parvesh Gupta	Government College, Saha
58	Deepak Kumar	G.M.N. College, Ambala Cantt.
59	Aarti Arora	Government College, Ambala Cantt.
60	Ajay Kumar	Department of Zoology, Kurukshetra University, Kurukshetra
61	Shashi Kant	Government College, Ambala Cantt.
62	Saddam Husain	Department of Zoology, Kurukshetra University, Kurukshetra
63	Jai Pal	S.D. College, Ambala Cantt.
64	Sunita	A.M. Sr. Sec. School, Kaithal
65	Bhawarth Sangwan	Engineer, J.B.C.C. Siwani
66	Anup Singh Sangwan	J.L.N. Government College, Faridabad
67	Vinod Kumar	Government College, Ambala Cantt.
68	Bhawarth Sangwan	Engineer, J.B.C.C. Siwani
69	Usha Rani Chahal	Government College, Ambala Cantt.
70	Nisha	Government College, Ambala Cantt.
71	Prabjot Kaur	Government College, Ambala Cantt.
72	Richa Lomas	Government College, Ambala Cantt.
73	Sujata Sasan	Government College, Ambala Cantt.
74	Anil Kumar	Government College, Ambala Cantt.
75	Darshan Lal	Government College, Ambala Cantt.
76	Nupur Sharma	Government College, Ambala Cantt.
77	Des Raj Bajwa	Government College, Ambala Cantt.
78	Bhavya Gandhi	Government College, Ambala Cantt.
79	Upendr Sethi	Government College, Ambala Cantt.
80	Satish Kumar Garg	Government College, Ambala Cantt.
81	Suresh Kumar	M.N. College, Shahabad

82	Nitika	Multani Mal Modi College, Patiala
83	Jyoti Shah	Multani Mal Modi College, Patiala
84	Sanjeev Kumar	Multani Mal Modi College, Patiala
85	Mandeep Kumar	Government College, Ambala Cantt.
86	Rajeev Sharma	Multani Mal Modi College, Patiala
87	Rohit Sachdeva	Multani Mal Modi College, Patiala
88	Sumeet Kumar	Multani Mal Modi College, Patiala
89	Sapna Dhiman	Multani Mal Modi College, Patiala
90	Varun Jain	Multani Mal Modi College, Patiala
91	Neeraj goyal	Multani Mal Modi College, Patiala
92	Sandeep Kaur	Government College, Ambala Cantt.
93	Seema	Government College, Ambala Cantt.
94	Harish Kumar	Government College, Ambala Cantt.
95	Vikas	Government College, Ambala Cantt.
96	Harish Kumar Soni	Government College, Ambala Cantt.
97	Komal Saroha	Government College, Ambala Cantt.
98	Anju Singla	Government College, Ambala Cantt.
99	Monika Mukul	Government College, Ambala Cantt.
100	Neeraj	Government College, Ambala Cantt.
101	Subhash Nautiyal	Government College, Ambala Cantt.
102	Anil Tomar	S.A. Jain College, Ambala City
103	Indresh Aggarwal	Government College, Ambala Cantt.
104	Indu Bala	Government College, Ambala Cantt.
105	Varsh Rani	Sai Nath university, Ranchi
106	Roopa Gupta	Government College, Ambala Cantt.
107	Amarjeet Kaur	Government College, Ambala Cantt.
108	Ajit Singh	Government College, Ambala Cantt.
109	Amit Pahwa	Government College, Ambala Cantt.
110	Himshikha Rahi	Punjabi University, Patiala
111	Avtar singh Rahi	Government College, Ambala Cantt.
112	Suyash Garima	Ambala City
113	Rohini Singh	Government College, Ambala Cantt.
114	Sandeep Singh	Government College, Ambala Cantt.
115	Meenakshi Tomar	Government College, Ambala Cantt.
116	Gurdev Singh Dev	Government College, Ambala Cantt.
117	Nidhi Sharma	Government College, Karnal
118	Rajni Saini	Government College, Ambala Cantt.

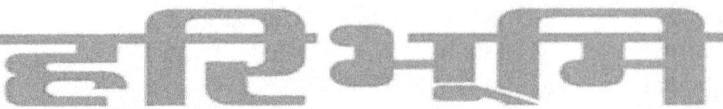

हरिभूमि

राष्ट्रीय हिन्दी दैनिक

कार्यक्रम | गवर्नमेंट पीजी कॉलेज में हुआ केमेस्ट्री पर नेशनल सेमिनार पर बोले: सिंह

विज्ञान व तकनीक से युग की पहचान

हरिभूमि न्यूज.अंबाला

अंबाला-पंचकूला के पुलिस कमिश्नर ओपी सिंह ने कहा कि कोई भी युग विज्ञान व तकनीक के युग के तौर पर जाना जाता है। उन्होंने कहा कि लोगों ने भले ही भगवान को नहीं देखा हो,लेकिन वैज्ञानिकों ने भगवान जैसा काम किया है। पुलिस कमिश्नर सिंह आज अंबाला छावनी के गवर्नमेंट पीजी कॉलेज में केमेस्ट्री विभाग की तरफ से एक दिवसीय राष्ट्रीय सेमिनार के उद्घाटन सत्र में मुख्य अतिथि के तौर पर बोल रहे थे। उन्होंने कहा कि वैज्ञानिक खोजों को बढ़ावा देने के लिए फंड और माहौल दोनों मुहैया करवाए जाने चाहिए। भले ही आज विज्ञान हर समस्या का हल नहीं ढूंढ पाया है। लेकिन विज्ञान व वैज्ञानिकों ने लोगों में आशा पैदा की है। सेमिनार में

अंबाला। सेमिनार को संबोधित करते पुलिस कमिश्नर ओपी सिंह।   फोटो: हरिभूमि

नोट स्पीकर व आईआईटी नई दिल्ली के केमेस्ट्री विभाग के अध्यक्ष प्रो.रविशंकर ने पर्यावरण प्रदूषण व उसके केमिकल व बायोलॉजिकल कारणों पर प्रकाश डाला। भले ही वैज्ञानिक खोजों ने इंसान का जीवन सरल बना दिया हो। इन खोजों ने

कई तरह के प्रदूषण को बढ़ावा दिया है। व प्रदूषण कम करना आज वैज्ञानिकों के लिए बड़ी चुनौती है। सेमिनार के रिसोर्स पर्सन व चितकारा युनिवर्सिटी पंजाब के डीन डा.इंद्रजीत सिंह संधू ने पर्यावरण प्रदूषण रोकने में नैनो टेक्नोलॉजी की

■ **वैज्ञानिक खोजों ने इंसान का जीवन सरल बना दिया**

भूमिका पर प्रकाश डाला। इससे पहले तीनों अतिथियों के कॉलेज पहुंचने पर कॉलेज प्राचार्या डा.कमलेश ने उनका स्वागत किया।

केमेस्ट्री विभागाध्यक्ष व सेमिनार के कॉर्डिनेटर डा.अवतार सिंह राही ने अतिथियों का परिचय करवाया। प्रो.गुरदेव ने मंच संचालन किया। सेमिनार के अंत में डा.रोहिणी ने अतिथियों का धन्यवाद किया व सेमिनार के समापन भाषण में पंजाबी युनिवर्सिटी के प्रोफेसर एन. एस. अत्री व सीनियर प्रोफेसर डा.सुबोध शर्मा,डा.रमेश मेहरा, डा.देशराज बाजवा,प्रो.सतीश, डा.दर्शन आदि का विशेष योगदान रहा।

दैनिक जागरण

World's Largest Read Daily

कॉलेज में समारोह के दौरान मौजूद आयुक्त।   जागरण

ISBN-13: 978-1533168078
ISBN-10: 1533168075

Chemistry of Industrial Globalization, Environmental Pollution and its Chem-Biological Significance February 2016

Industrial Globalization and Environmental Awareness

Proceedings of the NATIONAL SEMINAR held by

Department of Chemistry, Government PG College, Ambala Cantt - Haryana - India

INDEX

SrNo	Name	Topic	Pages
		From the Desk of Seminar Convener	7
		Details of Delegates participated	9
		Newspapers Cuttings	12
		Index	13
1.	Anshul Bansal, Sumita, Rashmi Dhawan and Anil K. Tomar	Global Warming and Greenhouse Effect: Concept, Consequences and remediation	15
2.	Bhawarth Sangwan and Smita Raghuvanshi	Biodegradation of Chlorinated Volatile Organic Compounds (VOCs) using mixed culture: A case-study on Carbon tetrachloride (CCl_4)	21
3.	Prerna Goyal and Nikhil Khurana	Effects of Chemical and Other Types of Industry on Health	29
4.	Nitika Arora	Impact of ICT on Environment	35
5.	Anup Singh Sangwan and Bhawarth Sangwan	Impact of Globalization on Industrial development	41
6.	Rashmi Dhawan, Sandeep Kumar and Anshul Bansal	Activated Carbon Adsorption Technology for VOC Emission Control	46
7.	Dr Deepak Sharma	Green Approach in The Synthesis of Heterocyclic Compounds from α-Tosyloxyketones and α,β-Ditosyloxyketones	51
8.	Dr Ramesh Sandhu	The Environmental Impact of E-waste	57
9.	Yogita Maheshwary	Oxidative Coupling of Tetrahydroisoquinoline with Alkynes	63
10.	Harish Kumar Soni and Parvesh Gupta	A Brief Review on Green Chemistry	68
11.	Rajeev Sharma, Jyoti Shah and Sanjeev Kumar	Indoor Pollutants-A threat to Human Health	81
12.	Himshikha Rahi	A Survey of Encryption Techniques to Secure Cloud Storage	87
13.	Veenu Saini	Green IT - Go for Green Gadgets	93
14.	Dr Rohini Singh and Dr Sheesh.P. Singh	The Effects and Implications of Climate Change on Plant Phenology	98
15.	Nisha Saini	Green Computing: Heading Towards a Better Future	105
16.	Jai Pal	Synthesis & Catalysis of Polystyrene-Anchored U(Vi) & Mo(Vi) Complexes	109
17.	Dr Avtar Singh Rahi	Effects of Toxic Chemicals on Environmental and Human Health	112
18.	Dr Ashima	Water Pollution from Pulp and Paper Mills	113

19.	Harish Soni and Deepak Kumar	Cause and Effects of Global Warming	114
20.	Pooja Sharma	Role of Chemistry in Environmental Protection	115
21.	Dr Alka	Current Scenario of E-waste Management in India	116
22.	Dr Anil Kumar	Volumetric and Ultrasonic Studies of Binary Liquid Mixtures of Alkoxypropanols with Cyclic Amide at Different Temperatures	117
23.	Vinod Kumar, Kuldeep Singh, Devinder Kumar and Mayank Kinger	Synthesis of Some Novel N-aryl- 2-mecrcaptoimidazoles as Potential Antimicrobial Agents	118
24.	Sushil Kumar & Sanjiv Arora	Physico-Chemical Studies of Nonlinear Optical Guest-Host Polymeric Thin Films	119
25.	Reena and Anil Kumar Anil Kumar and Reena	Recycling and the Environment Role of needs to explore different applications of Technology	120
26.	Parisha	Global Warming and Its Solutions	121
27.	Praveen Kumar Yadav	Clutter Processing for Highly Sensitive Radar Using Wavelet Transforms	122
28.	Dr Anil Jindal and Dr Surender Singh	Biodiversity Gain with Declining Population Sizes	123
29.	Maninder Kaur	Quark Diagram Analysis of Bottom Meson Decays Emitting Two Pseudoscalar Mesons	124
30.	Dr. Neeraj Goyal and Dr. Varun Jain	Corporate Sustainability Auditing – A Look Into Indian Corporate Sector	125
31.	Gulshan Singh , Ranjana Aggarwal	Synthesis of 7-amino- 2,5-diarylpyrazolo[1,5-a]pyrimidines for Antimicrobial Evaluation	126
32.	Shashi Sharma and Latika	Recycling: Use and Impacts	127
33.	Dr Rajendra Swain and Narinder Anchal	Global Warming: Causes and Effects	128
34.	Shilpy Aggarwal and Deepika Saini	Design and Synthesis of Some New Curcumin Analogs and Their Pyrazole Derivatives of Potential Biological Interest	129
35.	J.P. Deshwal, B.R. Deshwal, J P Saharan	Physio-chemical studies of water-quality parameters of drinking water-"A case study of Jind City, Haryana (India)"	130
36	Neha, B.R. Deshwal, S P Sharma, P. Singh & J P Deshwal	Studies of Quality Parameters of Drinking Water-"A Case Study of Rural Areas of Kaithal (Haryana)"	131
37	Joginder	Spectrophotometric Determination of W(Vi) After Extraction of Its Fhb Complex into Chloroform Solvent	132

Photographs: Moments of Sharing Time and Happiness

ISBN-13: 978-1533168078
ISBN-10: 1533168075

Chemistry of Industrial Globalization, Environmental Pollution and its Chem-Biological Significance

February 2016

Industrial Globalization and Environmental Awareness

Proceedings of the NATIONAL SEMINAR held by

Department of Chemistry, Government PG College, Ambala Cantt - Haryana - India

Global Warming and Greenhouse Effect: Concept, Consequences and remediation

Anshul Bansal[1*], Sumita[2], Rashmi Dhawan[1] and Anil K. Tomar[3]

[1]Department of Chemistry, S. A. Jain (PG) College, Ambala City-134002
[2]Department of Mathematics, S. A. Jain (PG) College, Ambala City-134002
[3]Department of Physics, S. A. Jain (PG) College, Ambala City-134002
*Corresponding author E-mail: anshulbansal001@gmail.com

Abstract

The climate of earth has always been, and still is, constantly changing. Greenhouse effect is a warming of earth's surface by a complex process involving sunlight, trace gases and particles in the atmosphere. The greenhouse effect is a natural phenomenon and vital to life. Without the greenhouse effect the earth's average temperature would be $-18^{o}C$, instead of current average temperature of $15^{o}C$. However problems may arise when the atmospheric concentration of greenhouse gases such as CO_2, CH_4, N_2O, O_3, halocarbons and water vapours increase. The recent IPCC (Intergovernmental Panel on Climate Change) Scientific Assessment of Climate Change estimated that continued accumulation of greenhouse gases may lead to a number of serious consequences like Global warming and climate change, rising sea level, disruption of the water cycle, worsening health effects, changing forest and natural ecosystem, challenge to agriculture and the food supply. In this connection many measures are to be taken, which would require improved technology and additional cost. There is an additional requirement to switch over to low or no carbon fuel and also to non-fossil fuel sources of energy such as solar, hydroelectric, nuclear etc. Production of CFCs, halons, carbon tetrachloride in refrigerators, air conditioners, solvents etc. should be banned. Alternative chemicals need to be developed. Recycling of chemicals will be required. It is advised to increase afforestation through agro forestry, social forestry, etc. and to decrease deforestation. The IPCC also proposed the policy of the Clean Development Mechanism (CDM) for joint implementation for credit in developing countries.

Introduction

Most climate scientists agree the main cause of the current global warming trend is human expansion of the "greenhouse effect"[1,2] — warming that result when the atmosphere traps heat radiating from Earth toward space. Climate change is a shift in the "average weather" that a given region experiences. Global climate change means change in the climate of the earth as a whole. Global climate change occurs naturally and the earth's natural climate has always been, and still is, constantly changing. The climate change we see today differs from previous climate change in both its rate and its magnitude.

Greenhouse effect: Concept

Greenhouse effect is a warming of the lower atmosphere (troposphere) and earth's surface by a complex process involving sunlight, trace gases and particles in the atmosphere. When electromagnetic radiations of shorter wavelength (UV, Visible and IR) from the sun enter the earth's atmosphere, about a third of it is reflected back to space[3]. Of the rest some is absorbed by the atmosphere, but most of it is absorbed by the surface of the earth. The earth emits energy at longer wavelengths, mainly middle or high infrared (IR) portion of the electromagnetic spectrum. Here the earth acts like a black body. According to the nature of a black body, the earth absorbs all the incoming radiations and reradiates energy at longer wavelengths to the outer space. Therefore, like a black body, earth must be radiating as much as the radiation absorbed i.e. the power radiated by the earth will be equal to the power received by the earth.[4]

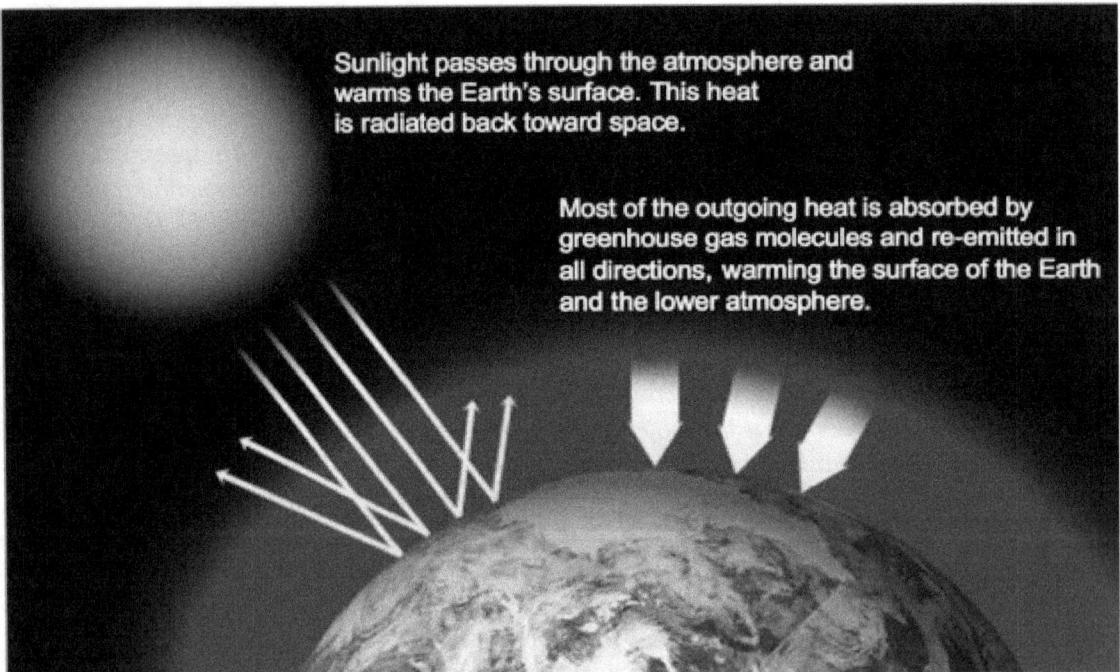

Figure1 Showing greenhouse effect (Picture taken from http://climate.nasa.gov/causes/)

As the sun is very hot it gives off its energy mainly in short-wave radiation (UV, Visible and IR). Most of this radiation passes through the earth's atmosphere but the energy radiated by earth is in the form of long-wave radiation which is much more easily absorbed than the short-wave radiation by the earth's atmosphere. Hence, the surface temperature of earth increases[5]. Some of this energy escapes to the space but much of this is absorbed by the greenhouse gases (CO_2, CH_4, N_2O, O_3, halocarbons and water vapours) in the atmosphere. Thus the earth's atmosphere heats up. This is natural greenhouse effect (**Figure 1**).

Greenhouse gases-Their Sources, relative contributions and effectiveness

Many greenhouse gases (GHSs) occur naturally. However, modern industry and lifestyles have led to new sources of greenhouse gases, as well as to the emission of entirely new greenhouse gases. The most important greenhouse gases are: Carbon dioxide, methane, nitrous oxide halocarbons, ozone and water vapour. The following properties are required for gases to become greenhouse gases[6]:

1. Dipolic nature
2. Having extremely strong, broad absorption band that overlaps wioth some of the wavelengths of heat radiations
3. Transition between the vibrational energy states
4. Formation of vibration molecular spectrum
5. Formation of vibration-rotation spectrum

The relative contribution and comparative effectiveness of greenhouse gases have been given in **Table 1**.

Table 1 Relative contribution, comparative effectiveness of greenhouse gases and their sources

Greenhouse gas	Relative contribution	Effectiveness compared to CO_2	Life span in atoms (Yrs.)	Sources
CO_2	50	1	100	Fossil fuel combustion, deforestation, change in land use, biomass burning, erosion
CH_4	18	30	10	Anaerobic bacterial activity in swampy lands, coal mines, gas leaks
N_2O	06	270	170	Fertilizer, biomass burning, fossil fuel combustion
O_3	12	2000	Short	Photochemical reaction
Halocarbons	14	CFC-11: 4300 CFC-12: 7100	70-170	Refrigerators, air conditioners

Consequences of global warming

The United Nations (UN) Environment programme established the Intergovernmental Panel on Climate Change (IPCC), with a remit to search for an authoritative international consensus of scientific opinion on climate change, its impact and possible responses. The IPCC's first assessment report in 1990 concluded that continued accumulation of greenhouse gases in the atmosphere would lead to climate change and were likely to have important impacts on natural and human systems. This was confirmed in second assessment report, published in 1996. The recent IPCC scientific assessment of climate change estimated the globally averaged surface temperature and these changes may lead to a number of serious consequences. These include:

a) Global warming and climate change: Global warming means surface temperatures have increased 0.6-1.2°F since the date 19th century. The 10 warmest years in this century all have occurred in the last 15 years. Of these, 2005 was the warmest year on record after 1850. The global average surface temperature rose 0.6 to 0.9 degrees Celsius (1.1 to 1.6° F) between 1906 and 2005, and the *rate* of temperature increase has nearly doubled in the last 50 years. Temperatures are certain to go up further (**Figure 2**).

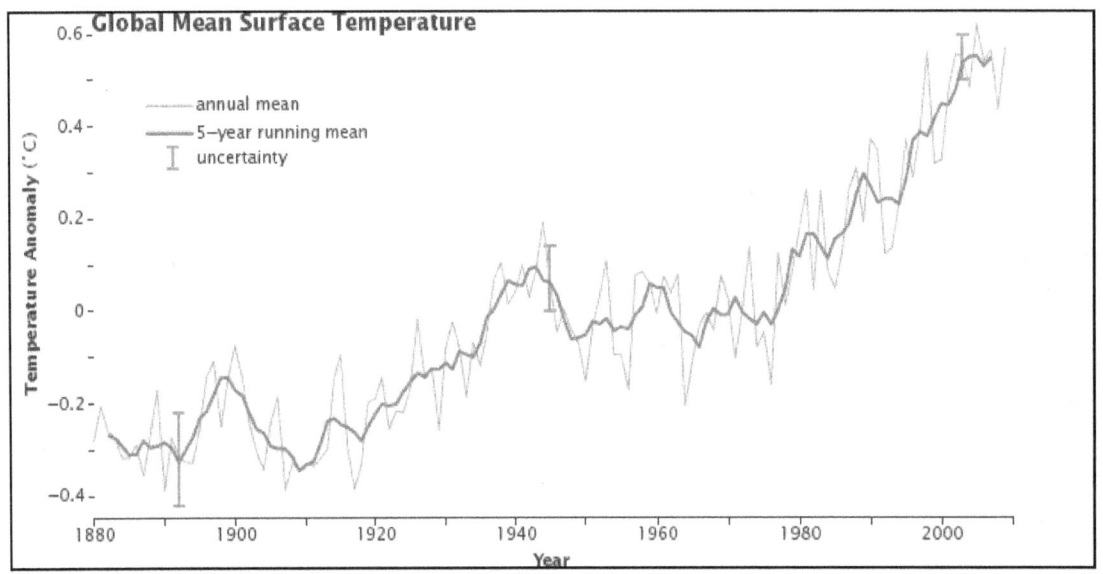

Figure 2 Showing Global average surface temperature is rising (NASA figure from Goddard Institute for Space Studies Surface Temperature Analysis.)

b) Rising sea level: Globally, sea level has risen 4-6 inches over the past century. In the next century, rapid global warming could triple that rate. By 2050, the oceans may rise another 8 inches, causing low-lying shorelines to recede significantly. The low-lying countries are the most vulnerable: Bangladesh, Indonesia, Pakistan, Thailand, Maldives, Egypt, Suriname and peninsular India.

c) Disruption of the water cycle: Evaporation will increase as the climate warms, which will increase average global precipitation. Soil moisture is likely to decline in many regions and intense rainstorms are likely to become more frequent that may lead to floods.

d) Worsening health effects: Climate change is likely to have wide-ranging and adverse impacts on human health. The projected increase in the duration and frequency of heat waves is expected to increase mortality rates as a result of heat stress. The heat in some hot weather regions may become unbearable, forcing people to migrate.

Climate change is also expected to lead an increase in the potential transmission of many infectious diseases, including malaria, dengue and yellow fever, extending the range of organisms such as insects that carry these diseases into the temperate zone.

e) Changing forest and natural ecosystem: A rapid and large scale climate change could severely harm the earth's ecosystem and such a change could make it difficult for many species to survive in the regions they now inhibit. Some could be forced to migrate, while others could become extinct.

f) Challenge to agriculture and the food supply: Regional changes in crop yields and productivity are expected to occur in response to climate change. There is likely to be an increased risk of feminine particularly in sub-tropical and tropical semi-arid and arid locations.

Steps to be taken to reduce emission of greenhouse gases

It is almost difficult, but not impossible to reduce the emissions of major greenhouse gases. As per provision of Kyoto protocol in 1997, there should be legally binding limits on the six major greenhouse gases (CO_2, CH_4, N_2O, O_3, halocarbons and SF_6) emissions of developed countries to an aggregate reduction of 5% on 1990 levels. In this regard, many steps were to be needed which would require improved technology and additional cost. The following measures may be adopted to check global warming[7,8]:

a) To increase energy efficiency in consumption and production. There is requirement to switch over to low or no carbon fuel.

b) To ensure complete combustion of fuel in vehicles through proper maintenance and to increase fuel efficiency. Cleaner burning fuel is required.

c) Shift to non-fossil fuel sources of energy such as solar energy, hydroelectric, nuclear etc.

d) Toxic emissions must be reduced. Production of CFC's, halons, carbon tetrachloride should be banned. Alternative chemicals will need to be developed.

e) To increase afforestation through agro forestry, social forestry, etc. and to decrease deforestation.

f) It has been found that plankton living in the surface water of oceans can control the global climate by up taking CO_2 through photosynthesis. The growth of plankton can be initiated by supply of iron and as a result greater the removal of CO_2 from atmosphere. Thus steps should be taken to promote the growth of plankton.

g) Kyoto protocol has proposed the policy of Clean Development Mechanism (CDM) for joint implementation for credit in developing countries. CDM will allow companies in developing countries to enter into co-operative projects to reduce emissions of greenhouse gases.

h) Recently, the United Nations Conference on sustainable development took place in Rio de Janeiro, Brazil on 20-22 June 2012 with an aim to ensure environmental protection especially from global warming.

To cope with the impact of global warming on our environment, it is right time for the world countries to come together and fight the problem in order to make our earth, a planet to live in, with a common goal of happy and prosperous life for all.

References

1. IPCC Fourth Assessment Report, 2007.

2. United States Global Change Research Program, "Global Climate Change Impacts in the United States," Cambridge University Press, 2009.

3. J. L. Chapman, M. J. Reiss, Ecology principles and applications, 1995, Cambridge University press, Cambridge.

4. F. B. John Mitchell, Reviews of Geophysics, 27, 1, 115-139, 1989.

5. A. K. Dey, Environmental Chemistry, 2001. New Age International Pvt. Ltd, New Delhi.

6. J. Gribbin, New Scientist, 22, 1-4, 1988.

7. M. Karpagam, Environmental Economics, 2001. Sterling Pub. Pvt. Ltd, New Delhi.

8. A. P. Sincero and G. A. Sincero, Environmental Engineering- a design approach, 1996. Prentice Hall, New Delhi.

ISBN-13: 978-1533168078
ISBN-10: 1533168075

Chemistry of Industrial Globalization, Environmental Pollution and its Chem-Biological Significance

February 2016

Industrial Globalization and Environmental Awareness

Proceedings of the NATIONAL SEMINAR held by
Department of Chemistry, Government PG College, Ambala Cantt - Haryana - India

Biodegradation of Chlorinated Volatile Organic Compounds (VOCs) using mixed culture: A case-study on Carbon tetrachloride (CCl$_4$)

Bhawarth Sangwan[1*], Smita Raghuvanshi[2]

[1*] *Process Engineer, Jai Bharat Gum & Chemicals Ltd., Siwani, Haryana, India*
Email: bhawarth@gmail.com
[2]*Assistant Professor, Department of Chemical Engineering, BITS, Pilani, India*
Email: smita@pilani.bits-pilani.ac.in

ABSTRACT

Carbon tetrachloride (CCl$_4$) is an important precursor of refrigerants (CFCs) but it is also a major constituent of GHGs (Green House Gases) and is responsible for Ozone layer depletion. The ubiquitous presence of Carbon tetrachloride (CCl$_4$) in earth's atmosphere is believed to originate exclusively from anthropogenic emissions. In the recent years, biological route has been identified as one of the most viable options available for the removal of chlorinated Volatile Organic Compounds (VOCs) due to less energy demand (1/10th to 1/4th) as compared to destructive chemical technologies. The current study aims to carry out batch preparation and kinetic study of Carbon tetrachloride (CCl$_4$) for optimizing Biofiltration process. It is carried out using mixed culture (present in sludge from waste water treatment plant) which has chlorinated VOCs removal capabilities. Mixed culture, as the consortium of different engineered species, develops symbiotic relationship which helps each other not only to survive in harsh conditions but also curb down chlorinated volatile organic compounds. Mixed culture is found to be superior candidate than selective strain for bio-degradation of CCl$_4$. It shows a promising future to develop a low cost, compact size closed system for *in-situ* chlorinated VOCs mitigation.

Keywords: Biofilm, Acclimatization, Anthropogenic, Autoclave, Supernatant, Inoculum.

1. INTRODUCTION

Adoption of stricter emission policies and increasing cost of chemicals & disposal of hazardous wastes in recent years, have provided the impetus to the development and optimization of biological treatment methods. Biofiltration has emerged as one of the cost effective biological air pollution control technologies for treatment of Volatile Organic Compounds (VOCs) emitted from chemical and process industries (Burgess et al. 2001). Also, biological treatment is environmental friendly, treatment is performed at ambient temperatures and it doesn't generate any secondary waste streams. Biological waste air treatment techniques have started to become the method of choice in many instances for the control of low concentrations of odors, volatile organic compounds (VOCs) or hazardous air pollutants in large air streams. In recent years, there has been significant maturation of biological waste air treatment research.

2. LITERATURE REVIEW

The elimination of a gaseous pollutant in a bio-filter is the result of a complex combination of different physicochemical and biological phenomena (Kondo and Bajpai 1999). In bio-filtration, the compound, to be treated, is passed through a packed bed of biomass supported on suitable materials such as compost, peat, wood chips, inert materials, etc. When waste solution or gases pass through the reactor, target organic pollutants diffuse into the biofilm with subsequent biological oxidation into less harmful substances such as CO_2, H_2O, minerals, etc. A consortium of microbial populations are known to play an important role in this process but current understanding of the mechanisms and specific microbial enzymes involved is limited (Busca and Pistarino 2003). In general, natural biomass support materials provide sufficient inorganic nutrients for micro-organisms. The required moisture is provided by saturating influent stream before it enters the reactor and/or by supplying liquid water intermittently. (Chou and Shiu 1997).

The overall effectiveness of a bio-filter is largely governed by the properties and characteristics of the support medium, which include porosity, degree of compaction, water retention capabilities and the ability to host microbial populations. Other performance parameters (removal efficiency) also include microbial inoculation, pH, O_2 concentration, initial concentration of pollutants, temperature, moisture, nutrient content, etc. The removal efficiency may be improved by chemical modification of the filter media or genetic modification of micro-organisms (Srivastava and Majumdar 2008).

2.1 Process of Biofiltration

The pollutant vapors and oxygen are transported as humid air by forced convection. Interphase mass transfer occurs and provided that the bio-filter bed particles are small, interfacial equilibrium is achieved so that gas-phase resistance can be neglected. In the biofilm, simultaneous diffusion and biodegradation of the pollutants occur as a result of growing or resting micro-organisms. The carbon dioxide resulting from the oxidation diffuses back and is further transferred to the gas phase. In most instances, micro-organism emissions and nutrient leaching can be neglected so that bio-filters can be considered as closed systems with respect to nutrient balance. The `cryptic` growth of microorganisms is not the only means of recycling nutrients because higher organisms like protozoa are also present in bio-filters. Even if they do not contribute directly to the pollutant elimination` they are certainly essential for nutrients 'cycles in the system (Deshusses 1997).

2.2 Mechanism of Biodegradation

The biodegradation is carried out by a complex ecosystem of degraders, competitors and predators that are at least partially organized into a biofilm (Devinny et al. 1998). There are three main biological processes that can occur in a bio-filter are attachment of micro-organisms, growth of micro-organisms, and decay & detachment of micro-organisms. The mechanisms by which micro-organisms can attach and colonize on the surface of the filter media of a bio-filter are transportation of microorganisms, initial adhesion, firm attachment and colonization (Loosdrecht 1990). The transportation of microorganisms to the surface of the filter media is further controlled by four main processes – diffusion (Brownian motion), convection, sedimentation due to gravity, and active mobility of the micro-organisms (Shim et al. 2003). Immediately after reaching the surface, initial adhesion occurs which may be reversible or irreversible depending upon the total interaction energy, which is the sum of van der Waal's force and electrostatic force. The factors that influence the rate of substrate utilization within a biofilm are substrate mass transport to the biofilm, diffusion of the substrate into the biofilm, and utilization kinetics within the biofilm. Adsorption and biodegradation perform simultaneously in bio-filters to remove biodegradable and water soluble hazardous organic molecules (Walle and Chian 1977).

2.3 Kinetics of Biodegradation

There are five reasonably well defined phases of microbial growth in batch cultures: lag, exponential, declining growth rate, stationary, and death (Fogg and Thake 1987). Michaelis-Menten equation (based on theoretical considerations) or Monod relationship (based on empirical considerations) can be employed in the calculation of developed enzyme mediated reactions. However, for very high concentration of substrate and where the substrate inhibition starts taking place, Haldane equation can be used. Kinetics are generally zero-order at the surface of filter, and order increases as depth of bio-filter increases. The growth of microbial is proportional to the size of microbial population.

3. MATERIALS AND METHODS

Biodegradation of CCl_4 has been studied by preparing batch solution of mixed culture and isolating the microbes from it. Kinetic study has been done using the prepared batch solution and different concentration of CCl_4 solutions.

3.1 Batch preparation

The batch preparation involved preparing the Minimal Salt Medium (MSM), Glucose stock solution, collecting & preparing active sludge sample and incubation & enrichment of microbes from sludge sample.

3.1.1 Preparation of minimal salt medium (MSM): It is done by adding different salts in distilled water. The compositions of different salts used to prepare MSM for the bioremediation study for VOCs (in g/l) are K_2HPO_4, 0.8; KH_2PO_4, 0.2; $CaSO_4.2H_2O$, 0.05; $MgSO_4.7H_2O$, 0.5; $(NH_4)_2SO_4$, 1; $FeSO_4.7H_2O$, 0.01.

3.1.2 Preparation of Glucose stock solution: It is done by dissolving 10 g of Glucose in 100 ml of distilled water which yielded 10,000 ppm solution.

3.1.3 Collecting & preparing the activated sludge sample: Activated sludge is a rich source of thriving micro-organisms and it was obtained from the Effluent Treatment Plant (ETP) of BITS, Pilani (Pilani Campus). The sludge (50-100 ml) after mixing with equal amount of distilled water in a beaker, is allowed to settle for half an hour in order to remove supernatant. The bottom portion is then again mixed with equal amount of distilled water and finally its upper layer (rich in microbes) is taken for the study.

3.1.4 Incubation & enrichment: It is carried on for six days. On first day, 2 ml of microbial culture sample with 1 ml of Glucose solution is added to 100 ml of MSM. The beaker is then kept in Incubator-shaker maintained at 37°C and 100 rpm for 24 h. Same procedure is repeated for next five days by increasing the dosage of CCl_4 by 0.2 ml each day and decreasing the dosage of Glucose solution by 0.2 ml each day. While preparing the batch for each day, 2 ml of content is pipetted out from the previous day batch sample and injected into the current day batch preparation. This experimentation is also termed as the *acclimatization,* since we are providing the conducive conditions to the micro-organisms to get habituated to the new environment.

3.2 Isolation of the Microbes using Microbial plating:

After acclimatization of mixed culture for six days, the final batch sample is used for doing the microbial plating on the Agar media plated petri dishes. This batch sample contains those micro-organisms (trained mixed culture) which have survived the toxicity of CCl_4 and have the capability to biodegrade CCl_4.The microbial plating is done on petri dishes using agar medium. The composition of agar medium is (in g/l) peptone, 5; beef extract, 3; sodium chloride, 5; agar, 17. The agar media is poured in the petri dishes and if there is no contamination in the petri dishes (after 12 h), microbial plating can be done. Using nano-pipettes, one micro-liter (10^{-6} liter) of the solution from the sample is poured and evenly spread on petri dish using L-rod. The petri dishes are then kept at 37°C for 24 h. Afterwards, the plates are analyzed for microbial colonies using the help of Colonies Forming Units (CFUs). The rest of the acclimatized microbe culture solution is first analyzed using UV-Vis spectrophotometer and then filtered out to find out the dry biomass weight.

3.3 Kinetic Study

Three methods are used to examine the growth of biomass as described below.The kinetic study is carried out for 40 h. Ten samples of MSM are autoclaved at 130°C. 2 ml of the final culture (6th day culture) is added to all the ten samples and then kept in the incubator-shaker. Samples were taken out at fixed time intervals.

3.3.1 Biomass Calculation: It is based upon the dry weight of dead biomass collected from the trained culture solutions of different concentration of CCl_4. Fixed volume of solution is pored through filter paper. The filter paper is then dried in oven for around 15 min. at 100°C. The difference in weight of filter is used to calculate the biomass (in g/l).

3.3.2 Optical Density Measurement: It is done using UV-Vis Spectrophotometer. The absorbance peak for pure CCl_4 solution is 230.86 nm. All readings of the samples were taken on the same wavelength.

3.3.3 CFU Measurement: It is measured in terms of number of colonies for 150 ppm concentration of solution. The dilution factor of 10^6 is used.

4. RESULTS AND DISCUSSIONS

4.1 Dry Biomass weight calculation

Three runs at 50, 100 and 150 ppm concentration of CCl_4 solution, have been analyzed to calculate the dry biomass weight in each solution at fixed time intervals. The readings of experimental runs are plotted in Figs 1- 3 down below.

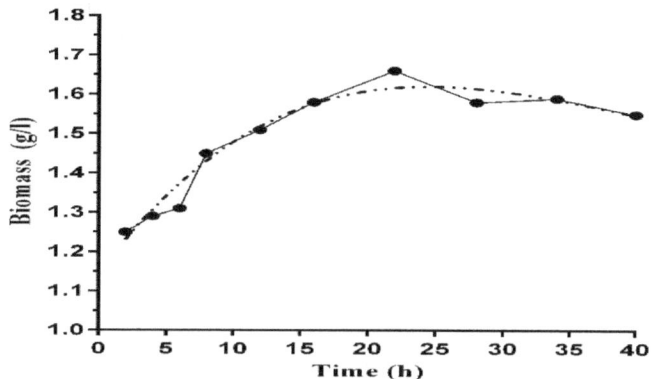

Fig. 1: Dry biomass weight for 50 ppm CCl₄ solution

The condition of the inoculum has a strong bearing on the duration of the *lag phase* (Spencer 1954) (marked by initial 8 h phase in Fig 1). An inoculum taken from a healthy exponentially growing culture is unlikely to have any lag phase when transferred to fresh medium under similar growth conditions of light, temperature and salinity. The growth rate of a micro-algal population is a measure of the increase in biomass over time and it is determined from the *exponential phase* (marked by 8 to 14 h phase in Fig 1). The duration of exponential phase in cultures depends upon the size of the inoculum, the growth rate and the capacity of the medium and culturing conditions to support algal growth. *Declining growth phase* (marked by 14 to 22 hour phase in Fig 1) normally occurs in cultures when either a specific requirement for cell division is limiting or something else is inhibiting reproduction. In this phase of growth, biomass is often very high and exhaustion of a nutrient salt, limiting carbon dioxide or light limitation become the primary causes of declining growth. Cultures enter *stationary phase* (marked by 22 to 28 h in Fig 1) when net growth is zero, and within a matter of hours cells may undergo dramatic biochemical changes. The nature of the changes depends upon the growth limiting factor. When vegetative cell metabolism can no longer be maintained, the *death phase* (marked by 28 to 40 h in Fig 1) of a culture is generally very rapid, hence the term "culture crash" is often used. Fig 1 following the trend proposed by Spencer et al. (1954) shows the increase in dry biomass with time for

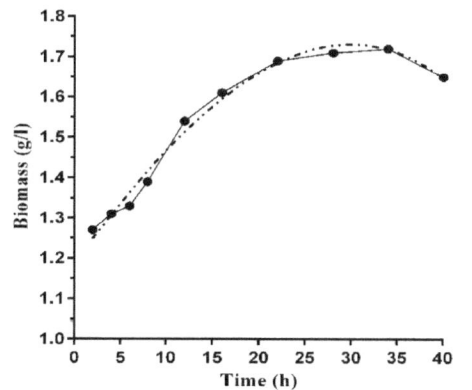

Fig. 2: Dry biomass weight for 100 ppm CCl₄ solution

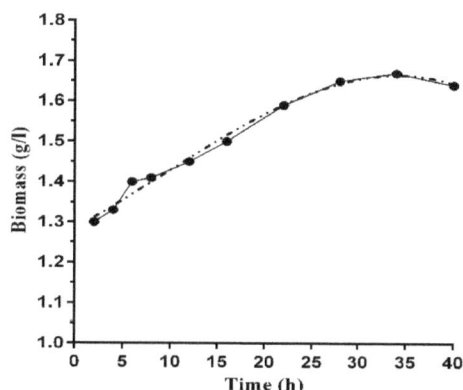

Fig. 3: Dry biomass weight of 150 ppm CCl₄ solution

most of the part of the study and then declines showing that culture has reached stationary phase. Fig 2 & 3 also confirms the lag, exponential, declining growth, stationary and death phase for the microbes for 100 and 150 ppm CCl_4 solution respectively. The pattern of curve remains the same for all three runs. However, it is imperative to note here that the amount of biomass present has decreased sharply in 150 ppm CCl_4 solution as compared to 50 and 100 ppm CCl_4 solution thus verifying the ineffectiveness of mixed culture to biodegrade the high concentration toxic solution.

4.2 Optical Density measurement

Fig 4 summarizes the plotted readings of spectrophotometer for 100 and 150 ppm CCl_4 solution.

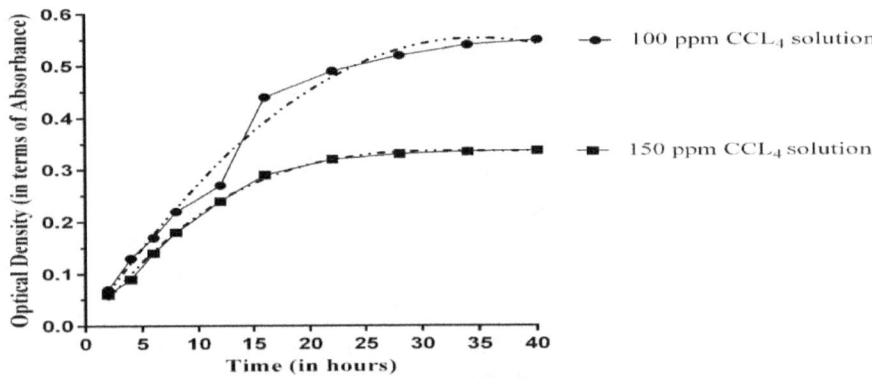

Fig. 4: Optical Density curve for 100 and 150 ppm CCl_4 solution

Fig 4 shows comparative analysis of biomass growth curve for 100 and 150 ppm CCl_4 solution. Optical Density increased with time in both the cases (100 and 150 ppm solution), however, growth and effectiveness of microbes for biodegradation is much more prominent in 100 ppm solution as compared to 150 ppm solution. Thus, once again verifying our assumption that mixed culture is ineffective for 150 ppm or above VOC solutions. Also, we can observe from Fig 4 that lag and exponential phase of both the curves are almost overlapping suggesting the similar microbial response in early phase of kinetic study.

4.3 Number of colonies

Colonies Forming Unit (CFU) is a direct observation method which involves counting number of colonies visually. CFU readings for 150 ppm CCl_4 solution are plotted in Fig 5 down below.

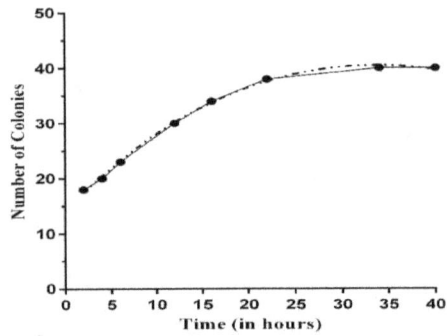

Fig. 5: Colonies Forming Units curve for 150 ppm CCl_4 solution

Fig 5 verifies our already concluded results about the five growth phases (lag, exponential, declining growth, stationary and death) of microbes as a reaction to chlorinated volatile organic compound. However, it is interesting to observe that curve failed to show any clearly demarked death phase ("culture crash") suggesting the incomplete growth curve. Thus, the experiment should be carried out for longer duration for higher concentration of CCl_4 solution in further studies.

5. CONCLUSION

The present study has shown that the concentration of biomass increases with time and then becomes constant (in fact slightly declines as well) showing that growth stops once the substrate is consumed completely. It also concludes that the microbes take some time to acclimatize to the new environment. The growth of the microbial culture follows the lag, log, declining, stationary and death phases. In future studies, analysis can be done using Gas Chromatograph for determination of exact concentration of CCl_4. Also, kinetic study can be extended to higher concentration of CCl_4 solution and for longer duration in order to clearly establish the microbial response curve for CCl_4 solution. Though, biodegradation efficiency will decrease sharply as the concentration of Carbon Tetrachloride increases, still it will be useful experiment to determine the exact behavior of microbial growth.

ACKNOWLEDGEMENT

Authors are thankful to Department of Science and Technology (DST), Govt. of India, for providing grant support (DST no.SR/FTP/ETA-07/2011) to the co-author Dr. Smita Raghuvanshi and Department of Chemical Engineering, BITS Pilani (Pilani Campus) for providing facilities to carry out research work.

REFERENCES

1. Altshuller, A. P., 1976. Average tropospheric concentration of carbon tetrachloride based on industrial production, usage, and emissions, *Environ. Sci. Technol.*, 10: 596-598.
2. Burgess, J.E., Parsons S.A., & Stuets R.M., 2001. Development in odors control and waste gas treatment biotechnology: A review, *Biotechnol. Adv.*, 19: 35-63.
3. Busca, G. & Pistarino, C., 2003. Abatement of ammonia and amines from waste gases: A summary. *J. Loss Prevent. Process Ind.*, 16: 157-163.
4. Chou, M.S. & Shiu, W.Z., 1997. Bioconversion of methylamine in bio-filters, *J. Air Waste Manage. Assoc.*, 47: 58-65.
5. Srivastava, N.K., & Majumdar, C.B., 2008. Novel Biofiltration methods for the treatment of heavy metals from industrial wastewater, *Journal of Hazardous Materials*, 151: 1–8.
6. Kondo, R., Bajpai, P., 1999. Biotechnology for environmental protection in the pulp and paper industry, *Springer, Germany.*
7. Deshusses, M.A., 1997. Biological waste air treatment in bio-filters, *Current opinion in Biotechnology*, 8: 335-339.
8. Devinny, J.S., et al., 1998. Bio-filtration of air pollution control, *CRC Press* 2:23.
9. Loosdrecht, M. C., 1990 Influence of Interfaces on Microbial Activity, *Microbial Reviews*, 54(1): 75.

10. Shim W.G., et al, 2003. Bio-filter in water and wastewater treatment, *Korean J. Chem. Eng.*, 20(6): 1054-1065.

11. De Walle, F., & E. Chian, E., 1977. Biological regeneration of powdered activated carbon added to activated sludge units, *Water Res.* 11:439–446.

12. Fogg, G. E., & Thake, B., 1987. Algae Cultures and Phytoplankton Ecology. 3rd edn, *London: The University of Wisconsin Press, Ltd*, p. 269.

13. Spencer, C. P., 1954. Studies on the culture of a marine diatom. *J. mar. biol. Ass. U.K.* 33: 265-290.

14. Charlotte, S., et al., 2011. Natural and enhanced anaerobic degradation of 1,1,1-trichloroethane and its degradation products in the subsurface: A critical review, water research 45(9), 2701 -2723.

15. Dennis, M.N., & Barford, J., 2000. Biofiltration as an odor abatement strategy, *Biochemical Engineering Journal* 5:231–242.

16. Bohn, H., 1992. Consider Biofiltration for decontaminating gases. *Chem Eng Prog* 88:34–40.

17. Raghuvanshi, S., & Babu, B.V., 2006. Removal of Methyl Ethyl Ketone (MEK) using Biofiltration, *Proceedings of National Conference on Environmental Conservation (NCEC-2006)*, BITS-Pilani, September 1-3, pp. 665-669.

18. Babu, B.V. and Raghuvanshi, S., 2003. Studies on Adsorption of Volatile Organic Compounds using Bio-filtration, *ME Thesis, Birla Institute of Technology & Science Pilani*.

ISBN-13: 978-1533168078
ISBN-10: 1533168075

Chemistry of Industrial Globalization, Environmental Pollution and its Chem-Biological Significance
February 2016

Industrial Globalization and Environmental Awareness

Proceedings of the NATIONAL SEMINAR held by

Department of Chemistry, Government PG College, Ambala Cantt - Haryana - India

Effects of Chemical and Other Types of Industry on Health

Prerna Goyal and Nikhil Khurana

Assistant Professor in Chemistry

M.M .P.G College Fatehabad, Haryana-125050

Email: nikhu4041@gmail.com

Abstract

The effects of chemical released from home, agriculture and various kinds of industries are very harmful. These chemicals came in contact with human body either by inhaling or in direct skin contact. Many disasters happened in previous years due to these chemicals. The outbreak of cadmium poisoning occurred in Toyama city of Japan in the form of Itai Itai and Ouch Ouch. Another disaster is due to mercury in 1935 in Japan by mixing of mercury in Minamata bay and this mercury enters into humans through the fishes of minamata bay. The effect of methyl iso cyanide is seen in Bhopal. Around 10,000 people were affected and their effects are still on the people. These effects cannot be removed totally but it can be minimized in a large extend by using organic farming, filtration of extraction gases etc.

Keywords: - Organic farming, Renewable sources

Introduction

We all are living in this world are based on only energy for our various purposes. Energy is of two types, mainly Renewable and non Renewable sources. Renewable sources are those which can be recovered back after sometime for example sun energy, wind energy but non renewable sources are those which we use for only one time in our life or their production needs thousands of years like petroleum, coal. To deals with our basic needs we mostly using non renewable sources but they causes pollution and leaves some waste products behind. As we move towards the 20th century, to develop ourselves and to meet population needs we moves towards globalization which leads to increase in industry and development , but along with it we have to suffer from many environmental degradation which causes a lot of adverse effect on human as well as animals health. So here we are discussing about the effects of chemical and other type of Industry on Health. These industries releases many type of waste, some wastes contain chemicals that are hazardous to people and the environment. Once these hazardous chemicals are present in the environment, people can become exposed to them. Exposure occurs when people have contact with a chemical, either directly or through another substance contaminated with a chemical. Chemicals can enter the environment from many different sources such as landfills, incinerators, tanks, drums, or factories. Human exposure to hazardous chemicals can occur at the source or the chemical could move to a place where people can come into contact with it. Chemicals can move through air, soil, and water. They can also be on plants or animals, and can get into the air we breathe, the food we eat and the water we drink. The different ways a person can come into contact with hazardous chemicals are called exposure pathways. There are three basic exposure pathways: inhalation, ingestion, and skin contact. Inhalation is breathing or inhaling into the lungs. Ingestion is taking something in by mouth. Skin contact occurs when something comes in direct contact with the skin. Ingestion can be a secondary exposure pathway after skin contact has occurred, if you put your hands in your mouth and transfer the chemical from your hands to your mouth.

Chemical and industry effects on human health

Health effects of chemical exposure:- we come into contact with many chemicals in our daily life, out of which some are useful and safe but others are the waste products and dangerous for our health. Harmful chemicals get into your body when you breathe, eat, drink or absorbed by our skin. People respond to chemical exposure in different ways. Some people may come into contact with a chemical and suffered from hazardous diseases, while some people in contact with same chemical never be harmed. The major factor that plays role in getting sick from a chemical are:-

a) The type of chemical you are exposed of in contact with chemical.
b) How much time you are in contact with that chemical
c) The quantity of chemical you are in contact or inhaled by(drinking, eating, smell)
d) The reaction of chemical with which part of body
e) The defense system of our body.
f) The frequency (how many times the person was exposed).

Types of Chemicals in kitchen

We commonly use different kind of chemicals in our daily life in kitchen. We are using a large no. of chemicals and directly taking it in our daily food products without our knowledge. These chemicals are so hazardous and they have a very large effect on our health. The common chemicals are used by adulterants present in our kitchen are: -

a) Milk: - In milk the common chemicals present are Urea, oxytocin, Detergents, Starch, and 40%formalin which is used as preservative.
   The harmful effects of urea:
   - The excess of urea increases the chances of weakness, stomach disorder and muscles cramps.
   - Urea decomposes to give NH_2 and CO_2 which causes irregular heart beat and chances of DNA breaking.
b) The harmful effects of formalin:
   - In stomach formalin produces many kind of gastritis which may leads to death.

- Dilute formalin can cause destruction upon pancreas, liver, fallopian tubes.

c) The harmful effects of oxytocin:
- It causes a large amount of weakness and headache.
- It causes tiredness and shortness of breath because of tightness in chest which leads to swelling around eyes, face, lips.

Types of chemicals from Agriculture

From last 25 years, to get more production and to get safe from insects and herbicide we are using a large no of fertilizers, pesticides and insecticides without any limit and control. These all consists of very dangerous chemicals which are commonly non biodegradable and moves from one cycle to another. The common in it are urea, D.D.T, B.H.C, D-trans Alithrin, Aldreine, Xylene, Organophosphate, Organochlorine, Carbamate, Heptachlor and chlordane etc.

Types of chemicals from mosquito pest control

According to the new research there are harmful chemicals present in our mosquito pest control sprays which we are using in our homes. The common chemicals are Dibrom-permethrin and D-trans Alithrin. It causes our genetic system, cancer causing and neuro-toxic dangers to unborn child.

Types of waste from Industry

Industries are very important for our daily life and are the backbone of country, but including profits it has many disadvantages. Industries generate a lot of waste products and polluted air. The pollutants usually consist of organic matter, inorganic dissolved salts, suspended solids, thermal constituents in form of heat and pathogens. If there is a high concentration of phosphate and nitrate, growth of aquatic plants will rise causing Eutrophication. Thermal pollution causes serious ecological problems.

Some harmful effects of different kind of Industries

1) Cement Industry: - In it the cement is prepared from calcium aluminate and calcium silicate of varying compositions. The pollution is caused by very fine particles of cement dust into environment. Its harmful effects are:-
   a) Cement dust is inhaled by human lungs and is responsible for lungs, bronchitis, respiratory and cardiovascular diseases
   b) A large amount of waste heat is released during preparation, which causes loss of energy, degradation of energy sources and causes a lot in Global warming.

2) Sugar Industry: - Sugar industry is the one of the largest sugar producing country in the world. The raw material is sugarcane and waste products are Molasses and bagasse. Its harmful effects are:-
   a) A large amount of flyash or bagasse is present in atmosphere which causes respiration problems.
   b) During processes huge amount of waste hulls, stalks and steeps are formed which causes foul smell.

3) Mining and Metallurgy Industries:- A large amount of ore is obtained in the mining process which is then purified to get the pure metal, in this process the waste

products are mainly sulphides and chlorides which effects our respiratory system and even causes cancer.

4) Polymer Industry:- These are the giant big molecules which is the synonym of plastic, it is very popular due to durability, flexibility and low cost but responsible for extensive degradation because of its Non biodegradability its harmful effects are:-
   a) Polyvinyl chloride is significant source of dioxin and secondary hazardous waste
   b) Dioxin is carcinogenic and a powerful hormone disrupter
   c) All phthalates (diethyl phthalate, benzyl butyl phthalate) used in food packaging, cosmetics, varnishes and plastic toys may cause birth defects, damage kidneys and liver.
   d) Breathing styrene form polystyrene and cause Leukemia
   e) Burning of polyethylene generates formaldehyde
   f) Incineration of plastic produces deadly poisonous gases like chlorinated dibenzofurans, CFCs, HCN, CO, SO_2 which causes harmful effects to body and depletes ozone layer.

5) Fertilizer Industry:- In these industry we prepare urea, Ammonium sulphate, Ammonium nitrate, super phosphates, etc. to increase the power of fields and more yield of crops. The waste products of industry are spill overs from manufacture of acids, boiler blow down and cooling water. Its harmful effects are:-
   a) Large amount of nitrogen , sulphides and ammonia is produced which causes eutrophication.
   b) High amount of fluoride is present in phosphatic fertilizer causing dental and skeletal flurosis to humans
   c) It also causes abnormal calcification of bones in animals and plants
   d) DDt, BHC, heptachlor are non biodegradable and their accumulation in human food chain.
   e) Higher concentration of DDT, BHC can cause anxiety, tension, cancer, mutations, stress etc.
   f) These all affect the vital organs, heart, brain kidneys and liver producing chronic disturbances

6) Electroplating Industry:- A ferrous or non ferrous material is electroplated with Ni, Cu, Zn, Cr, Pb, Cd, Al etc. to alter the surface properties of base metal. It consists of mainly three processes mainly surface cleaning by saponification with warm alkali, then treated with HCL to remove rust and then we do plating in electrolytic cells. The harmful effects are:-
   a) Electroplating effluents are highly toxic and corrosive.
   b) Cyanide, chromates and salts of heavy metals are toxic to aquatic life.

Different kinds of chemicals, their origin and effects:-

S.No.	Chemical	Origin	Effects
1	Radon	Ground	Lungs, cancer, decreases oxygen supply in blood, asbestosis
2	Cadmium	Old batteries, cigarette smoke, ores of zinc	Anemia, bones become fragile and brittle, cancer, damage to liver kidney and lungs

3	Benzene	Degreasers, laboratories	Respiratory, lungs, highly flammable, decreased oxygen supply in blood
4	Lead	Old paint, outdated plumbing	Stops hem-synthesis, vomiting, brain damage, anorexia, paralysis
5	Mercury	Thermostat, thermometer, fish	Minamata, loss in limbs, hearing, impaired vision, mental retardation damage of DNA
6	Uranium	Food and water, proximity to nuclear plants	Decrease formation of urine, kidney damage, decreased blood flow to kidney
7	Carbon disulphide	Industrial production	Affect cardiovascular system, heart failure, veins blockage
8	Nickel	Cement, electroplating industry	Skin diseases, irritation to eyes, redness, dermatitis
9	Chromium	Paints, industrial pollution, electroplating	Slows down of immune system, bone marrow, allergy
10	Methylene chloride	Paint removers, auto parts cleaning	Liver damage, tumors, death of liver cells, accumulation of fats
11	Volatile organic compounds	Fumes from gasoline,, paint, adhesives, building supply	Respiration problems, lungs cancer, skin irritation
12	Carcinogens	Plastic, fibers, dyes, pvc, paper and textile industry	Bladder cancer and lung cancer, suspected carcinogen
13	Methyl iso cyanide	Carbamate industry	Extremely hygroscopic, broncho pneumonia, cough, constriction, chest pain, deaths
14	Cyanide	In seeds of fruit, metal cleaning and electroplating	Form cyanide complex, unable to carry oxygen, death
15	Arsenic	From sea water, soils, arsenates	Uncoupling of phosphorylation by arsenic, sensory paralysis, bone marrow

Conclusion

According to WHO report around 7,50,000 people are poisoned by pesticides every year and around 14,000 deaths are caused due to chemical contamination. It is a big challenge to decrease the use of these harmful chemicals. The amount of pesticides on fruit and vegetables can be decreased by lots of washing and by doing organic farming. The central pollution control board should be strict against the pollution causing industries and the waste chemicals should be disposed off after treating chemicals. These harmful chemicals minimized up to a large extend but cannot be totally removed off.

Bibliography:-

- The Environmental chemistry by H. kaur

- www.chennaicorporation.gov.in
- http://avogadro.chem.iastate.org
- www.livestrong.com
- www.drugs.com

ISBN-13: 978-1533168078
ISBN-10: 1533168075

Chemistry of Industrial Globalization, Environmental Pollution and its Chem-Biological Significance

February 2016

Industrial Globalization and Environmental Awareness

Proceedings of the NATIONAL SEMINAR held by

Department of Chemistry, Government PG College, Ambala Cantt - Haryana - India

Impact of ICT on Environment

Nitika Arora

Assistant Professor in Computer Science

Govt. College For Women, Karnal

Nitikabhandari970@yahoo.com

Introduction

Information and Communication Technologies (ICTs) can improve environmental performance and address climate change across the economy. Both developed and developing countries face many environmental challenges, including climate change, improving energy efficiency and waste management, addressing air pollution, water quality and scarcity, and loss of natural habitats and biodiversity. The Internet and information and communication technologies (ICTs) are transformative technologies in that they put intelligence at the edges of networks, thereby maximizing users' capacity to create and adapt. Examples of such transformation include using ICTs to improve practices in agriculture and forestry; monitor air and water pollution; improve disaster warning and relief; improve the efficiency of the energy, transportation, and goods and services sectors; and harness social networking for transformative change. At the same time, the sustainability of these technologies must also be managed to avoid unintended consequences such as increased consumption and environmental damage from electronic waste. This paper explores how the Internet and the ICT and related research communities can help tackle environmental challenges in developing countries through more environmentally sustainable models of economic development, and examines the status of current and emerging environmentally friendly technologies, equipment and applications in supporting programs aimed at addressing climate change and improving energy efficiency. This paper provides an overview and points to examples of current activities and opportunities in each of these areas.

ICT and the knowledge-based economy

Traditional development models have focused on a shift from agriculture to manufacturing, the development of free markets, encouraging exports and industrialization in labour-intensive consumer goods – a model borne out in The East Asian Miracle (World Bank 1993) and the emergence of China as the World's largest exporter of ICT and related consumer equipment. Sheehan (2008) suggests a re-think, based on the evidence from the emergence of India and the thrust of China's Eleventh Five Year Plan (2006-11). Looking at long-term trends in employment and sectoral GDP shares and growth rates, Sheehan (2008) suggests that India provides an example of a "big-push" development driven by services (Figure 1), and that: "industrialization as it used to be understood is no longer a realistic option for most developing countries, and they need to find ways of participating in the growth of the modern services sector, which can directly improve the living standards of their people". It is notable that India's CO_2 intensity per unit of GDP is substantially lower than is typical of developing countries, comparable to that of Japan and lower than Germany's (Ghosh 2009).

Figure 1:   Value added shares by sector, India 1950–51 to 2007–08

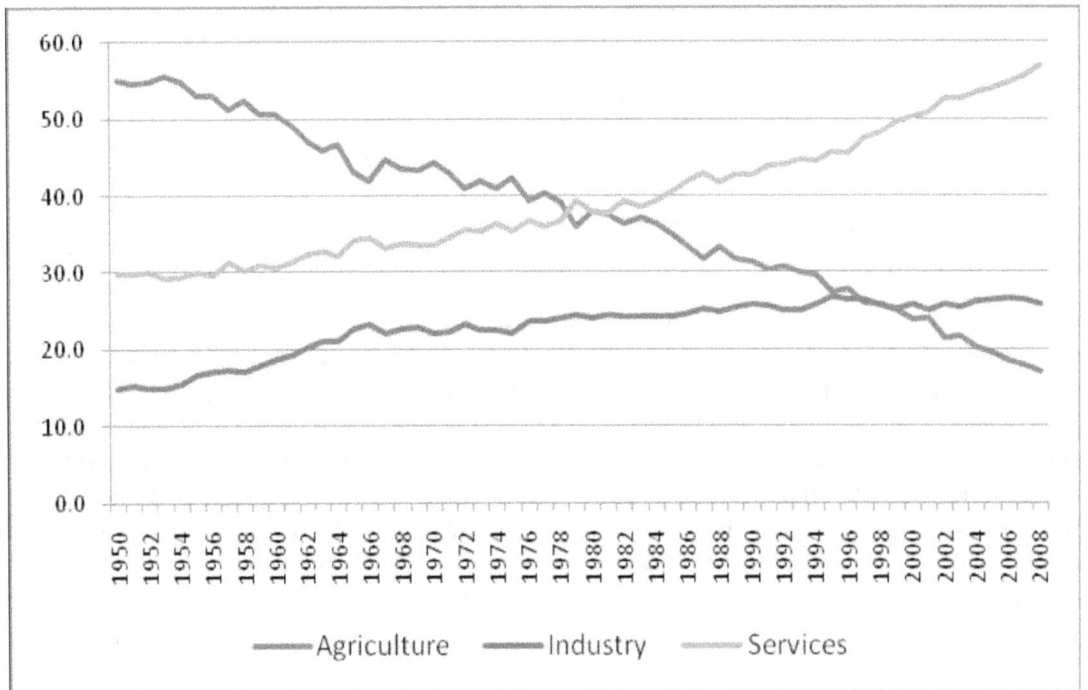

Source: Sheehan, P.J. (2008) *Beyond Industrialization: New Approaches to Development Strategy Based on the Services Sector*, UNU-WIDER Research Paper 2008/60: Helsinki.

ICTs have played a key role in making services tradable and the globalization of IT and IT-enabled services. Looking at the intensity of IT and IT enabled services exports, Houghton and Welsh (2009) note that in only four countries did computer and information services account for more than 25% of total services exports during 2006 – India, where they accounted for almost 40% (down from 50% in 2004), Ireland 31% (down from 39% in 2004), and Israel 27% . Their analysis suggests that IT and IT-enabled services exports can play an important role in a wide range of developed, emerging and developing economies, and may in the latter provide the basis for a more environmentally sustainable development path that has characterized industrialization in the past.

        Assessing the possibility of alternative development pathways, Berkhout *et al*. (2009) argue that the convergence of economic structures and growth rates, which plays such a central role in growth theories, does not imply that the emergence of socio-technical systems underpinning growth must also be convergent in terms of their technological composition and environmental quality, and call for greater attention to the resource and environmental quality of development as the basis of more sustainable development pathways.

ICT and the environment

The relationship between ICTs and the environment is complex and multifaceted, as ICTs can play both positive and negative roles. Positive impacts can come from dematerialization and online delivery, transport and travel substitution, a host of monitoring and management applications, greater energy efficiency in production and use, and product stewardship and recycling. Negative impacts can come from energy consumption and the materials used in the production and distribution of ICT equipment, energy consumption in use directly and for cooling, short product life cycles and e-waste, and exploitative applications (*e.g.* remote sensing for unsustainable over-fishing (Daly 2003).

The impacts of ICT on the environment can be direct (*i.e.* the impacts of ICTs themselves, such as energy consumption and e-waste), indirect (*i.e.* the impacts of ICT applications, such as intelligent transport systems, buildings and smart grids), or third-order and rebound (*i.e.* the impacts enabled by the direct or indirect use of ICTs, such as greater use of more energy efficient transport). Exactly what the impacts of ICT are, and to what extent there may be rebound effects (Box 1), are widely discussed topics. However, it is clear that attempts to measure the impacts of ICT on the environment should take account of the potential rebound effects and the entire life cycle, rather than simply the direct impacts of the product or application itself (Plepys 2002; Yi and Thomas 2007; Hilty 2008; etc.).

Estimates of the direct impacts of the ICT industries vary with the definition of the industry and coverage of ICT-related energy uses, but the production and use of ICT equipment is estimated to be equivalent to 1% to 3% of global CO_2 emissions (including embedded energy) and a higher and growing share of electricity use. In 2006, it was estimated that ICT equipment (excluding broadcasting) contributed around 2% to 2.5% of worldwide Greenhouse Gas (GHG) emissions – 40% of this was reported to be due to the energy requirements of PCs and monitors, 23% to data centres, 24% to fixed and mobile telecommunications, and 6% to printers (Kumar and Mieritz 2007). More recent life cycle assessments produce broadly similar results (Malmodin 2009). Data centres are a particular focus, and Koomey (2007) estimated that worldwide electricity use for servers doubled between 2000 and 2005, and he suggested that consumption would increase by a further 40% by 2010.

Figure 2: ICT Impact: The global footprint and the enabling effect

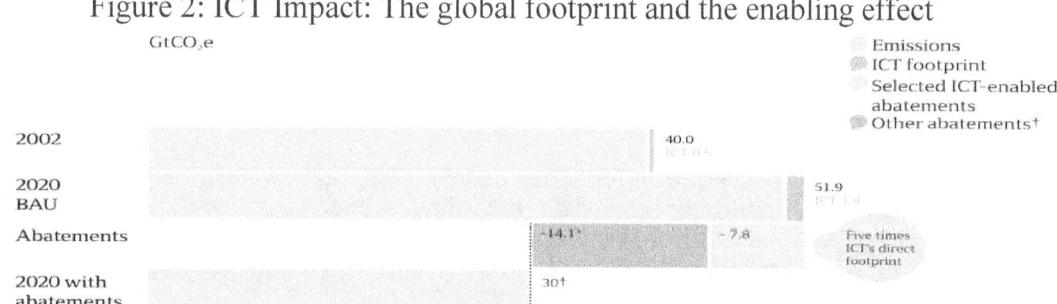

Source: The Climate Group (2008) *SMART 2020: Enabling the low carbon economy in the information age*, London, p15.

Nevertheless, the indirect enabling impacts of ICTs are greater, and a number of studies have identified potentially significant net positive impacts from ICTs. For example, The Climate Group (2008) identified key areas of enabling impacts potentially leading to global emissions reductions by 2020 that were five times the ICT sector's direct footprint (Figure 3). ICT and the Internet are enabling an increasing number of products and services to be delivered online (*i.e.* de-materialization). This affects scientific journals, books, music CDs, film and videos, software, etc., with fewer taking a physical form and less energy and potentially fewer resources being used in their production, storage and delivery. E-commerce and online shopping can save time and travel in searching and pricing, and centralized fulfillment and delivery can replace many thousands of individual trips, not only saving energy directly but also through potential reductions in traffic congestion. E-mail has replaced many millions of letters, written on paper, collected, sorted and delivered worldwide, with almost instantaneous communication that has a very small environmental footprint (Schmidt and Kloverpris 2009).

ICTs offer the potential for transport and travel substitution. With tele-work or e-work the reduction of transport and commuting time can be substantial and considerable benefits can accrue for individuals, employers and the community. The reduction of long distance travel possible as a result of the use of data, voice and video applications over IP for webcasts, tele-conferencing and video-conferencing can also be significant, and there are both direct impacts in terms of the environmental footprint and indirect impacts such as reduced demand on transport infrastructures and office facilities.

ICTs can also contribute to the resource and energy efficiency of many physical products embedded in either the products themselves or their production processes. For example, automotive electronics in the form of ignition chips have greatly improved the energy efficiency of motor vehicles, and industrial and household equipment and the design, construction and management of buildings increasingly includes "smart technology" to better control resource and energy use, emissions, serviceability and durability.

Nevertheless, there have been many studies pointing the difficulties in avoiding rebound effects and realizing the potential benefits (see Box 1), and it has been noted that the „paperless office" has not yet eventuated, e-commerce may not save energy if it encourages long distance delivery, tele-working can increase the home use of energy and demand for electronic equipment, such as routers and printers, and so on (Plepys 2002). As always, the key is not the technology, but how it is implemented and used.

Looking at ICTs as tools for dealing with environmental issues from a developing and emerging country perspective, ITU (2008) noted six application categories (Figure 4).

1) *Environmental observation*: terrestrial (earth, land, soil, water), ocean, climate and atmospheric monitoring and data recording technologies and systems (remote sensing, data collection and storage tools, telemetric systems, meteorological and climate related recording and monitoring system), as well as geographic information systems (GIS).
2) *Environmental analysis*: once environmental data have been collected and stored, various computational and processing tools are required to perform the analysis. This may include land, soil, water and atmospheric quality assessment tools,

including technologies for analysis of atmospheric conditions including GHG emissions and pollutants, and the tracking of both water quality and availability. The analysis of data may also include correlating raw observational data with second order environmental measures, such as biodiversity.

3) *Environmental planning*: at the international, regional and national level, planning makes use of the information from environmental analysis as part of the decision-making process for the purpose of policy formulation and planning. Planning activities may include classification of various environmental conditions for use in agriculture and forestry and other applied environmental sectors, and is often focused on specific issues such as protected areas, biodiversity, industrial pollution or GHG emissions. Planning may also include the anticipation of environmental conditions and emergency scenarios, such as climate change, man-made and natural disasters.

4) *Environmental management and protection*: involves everything related to managing and mitigating impacts on the environment as well as helping adapt to given environmental conditions. This includes resource and energy conservation and management systems, GHG emission management and reduction systems and controls, pollution control and management systems and related methodologies, including mitigating the ill effects of pollutants and man-made environmental hazards.

5) *Impact and mitigating effects of ICT utilization*: producing, using and disposing of ICTs require materials and energy and generate waste, including some toxic waste in the form of heavy metals. ICT use can mitigate the environmental impacts directly by increasing process efficiency and as a result of dematerialization, and indirectly by virtue of the secondary and tertiary effects resulting from ICT use on human activities, which in turn reduce the impact of humans on the environment.

Figure 3:   ICT application categories

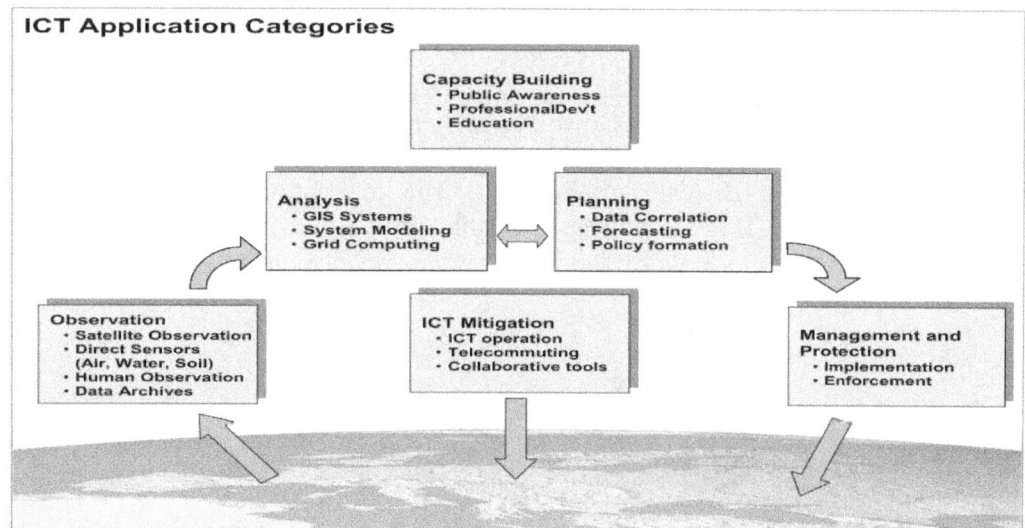

Source: ITU (2008) *ICTs for e-Environment: Guidelines for developing countries, with a focus on climate change*, ITU, Geneva, p25.

6) *Environmental capacity building*: efforts to improve environmental conditions rely on the actions of individuals and organizations. Capacity building includes efforts to increase public awareness of environmental issues and priorities, the development of professionals, and integrating environmental content into formal education.

Summary and conclusions

ICTs are all but ubiquitous and the potential uses and impacts of ICTs on the environment are many and varied. No short paper can cover all aspects, and this paper does no more than provide a few examples. However, it is possible to note some of the key areas of impact and potential in more general terms, highlighting some of the major issues arising for policy coherence.

References

1. Access Economics (2009) *The economic benefits of intelligent technologies,* Access Economics, Canberra. Available http://www.accesseconomics.com.au/
2. Barton, J.H. (2008) *Mitigating Climate Change Through Technology Transfer: Addressing the needs of developing countries*, EEDP Pare 08/02, Chatham House, London. Available http://www.chathamhouse.org.uk/files/12357_1008barton.pdf
3. Basel Convention (2003) *Report of the Conference of the Parties to the Basel Convention on the Control of Transboundary Movements of Hazardous Wastes and their Disposal,* UNEP. Available www.basel.int/meetings/cop/cop6/english/Report40e.pdf
4. Climate Risk (2008) *Towards a High-Bandwidth, Low-Carbon Future: Telecommunications-based opportunities to reduce greenhouse gas emissions,* Climate Risk, Sydney. Available www.climaterisk.com.au
5. Daly, J. (2003) *ICT and Ensuring Environmental Sustainability*, Development Gateway (dgCommunities). Available http://topics.developmentgateway.org
6. Houghton, J.W. and Welsh, A. (2009) *Australian ICT Trade Update 2009*, Australian Computer Society, Sydney. Available http://www.cfses.com
7. Koomey, J. (2007) *Estimating total power consumption by servers in the U.S. and the world,* Analytics Press, Oakland. Available http://enterprise.amd.com/Downloads/svrpwrusecompletefinal.pdf
8. Vetter, T. and Creech, H. (2008) *The ICT Sector and the Global Connectivity System: A sustainable development overview*, The International Institute for Sustainable Development, IISD, Winnipeg. Available www.iisd.or

ISBN-13: 978-1533168078
ISBN-10: 1533168075

Chemistry of Industrial Globalization, Environmental Pollution and its Chem-Biological Significance
February 2016

Industrial Globalization and Environmental Awareness
Proceedings of the NATIONAL SEMINAR held by
Department of Chemistry, Government PG College, Ambala Cantt - Haryana - India

Impact of Globalization on Industrial development

Anup Singh Sangwan[1*], Bhawarth Sangwan[2]

[1*]Associate Professor, Department of Economics, PJLN Government College, Faridabad
Email: anupsangwan64@gmail.com
[2]Process Engineer, Jai Bharat Gum & Chemicals Ltd., Siwani, Haryana, India
Email: bhawarth@gmail.com

ABSTRACT

Industrial development is of utmost significance to all types of economics. Industrial development is necessary for under-developed economies because they solve the myriad problems of general poverty, unemployment, income inequality, backwardness, low standard of living etc. On the contrary, industrial development is equally significant for developed countries as it helps to maintain their existing growth rate. The Government of India is going with many policy reforms after the severe fiscal crisis in 1990-91. After nearly more than two decades of globalization, one question constantly arises in the mid of economists that is, what has been the impact of globalization on industrial development of India? This paper analyses the positive and negative effects of globalization on industrial development. This analysis is based on secondary data of Index of Industrial Production (IIP).

Keywords: Index of Industrial Production (IIP), Mahalanobis model.

1. INTRODUCTION

Globalization can be defined, simply an expression of economic activities across political boundaries of nation states. More importantly, perhaps, it refers to a process of deepening of economic integration, increasing economic openness and growing economic interdependence between countries in the world economy. It is associated not only with a phenomenal spread and volume of cross border economic transactions, but also with an organization of economic activities which straddles internal boundaries. Globalization in India generally taken to mean integrating the economy of the country with the world economy by opening up the economy to foreign direct investment (FDI) by providing the facilities to foreign companies to invest in different fields of economic activities in India;

1. Removing constraints and obstacles to the entry of Multi-National Companies (MNCs).
2. Allowing Indian companies to enter into foreign collaboration in India and also encouraging them to set up joint venture abroad.
3. Bringing down the level of import duties.

However, the real thrust to the globalization process was provided was provided by the new economic policy integration introduced by the Government of India in July 1991 at the behest of International Monetary Fund (IMF) and the World Bank. Therefore, globalization has been identified with the policy reforms of 1991 and the subsequent extension of these reforms carried out in later years.

2. LITERATURE REVIEW

Globalization is an important issue. The major trends and performance of the economy are reflection of globalization process. Globalization, being significant for every economy of the world, has drawn interest of many scholars, academicians, social scientists and others to write on the subject. Although, it is very difficult to sort out the literature on globalization, but, there are mainly two schools of thought on this topic. First one argues, globalization is good for India and the second one counters the argument of first school of thought by reasoning the negative impacts of globalization for India.

First school of thought argued that trade without any barriers and competition is good for the whole world for providing better goods and services to the consumers.

Second school of thought is based on the impact of liberalization and opening up of market on Indian economy. Athreye (1999) and Kumar (2003) have studied the impact of foreign direct investment (FDI) on Indian economy. Sliglitz (2002) argues that globalization hasn't met its promises to the developing economies.

The objective is this paper can be summarized as – to analyze the impact of globalization on industrial development, to identify the areas where globalization has succeeded and to identify the areas where globalization has failed.

3. METHODOLOGY

The present study is based on secondary sources of data collected from various publications of government surveys, journals, books, magazines and reports prepared by scholars & economists. The reference period of the current study is post 1991 reforms (also known as New Industrial Policy of 1991) which may be considered as path towards globalization. The current work is an attempt to investigate the impact of globalization on industrial development of India.

4. RESULTS AND DISCUSSION

4.1 Variables of Globalization

Post-independence, the Indian economy was in doldrums due to decades of British rule and their ruthless policy to suck Indian economy dry of its all resources to facilitate their own growth. In 1947, the Indian economy took the path of socialism and started the process of industrialization with the Mahalanobis model of heavy industrialization, which was entirely state sponsored and regulated. The planned nature of Indian economy showed galloping growth in the first two plans but war with Pakistan and China along with recurring famine throughout the 1970's undermined the overall economic prosperity. It was in the 1980's when the need for economic reforms was felt from each sector of Indian economy. The reforms of 1991 in industrial, financial and external sector lead Indian economy on the path of globalization, where the role of state shifted from being protectorate to promoter of various sectors.

In the line with globalization, New Industrial Policy (1991) deregulated the industrial economy in a substantial manner. The major objectives of new policy were, "to build the gain already made, correct the weakness that might have crept in, maintain substantial growth in productivity & gainful employment and attain international competitiveness". In the pursuit of these objectives, the government announced a series of initiatives in the following areas;

(a) Abolition of Industrial Licensing – As of now, licensing is compulsory only for five industries. These are alcohol, cigarettes, hazardous chemicals, electronics, aerospace & defense equipment and industrial explosives.

(b) Public sector's role diluted – Now, only three industries are reserved exclusively for public sector – Atomic Energy, Minerals and Railways. Even in the field of railways, private investment in infrastructure development was allowed in 2014.

(c) Monopolistic and Restrictive Trade Practices (MRTP) Act limit removed – The Act has been accordingly amended giving more emphasis to the prevention and control of monopolies, restrictive and unfair trade practices. Henceforth, consumers are adequately protected from such practices.

(d) Relaxed norms for foreign investment and free entry to foreign technology – The main advantage of this assistance is that foreign investment also brings with it, the technical expertise, machine, and capital goods etc. which are scarce in underdeveloped countries.

4.2 Impact of Globalization

- Indian foreign exchange reserves which had fallen to barely $ 1 billion in June 1991. It rose substantially to over $ 353.46 billion in July 2015.
- Liberalization and openness have actually increased over self-reliance. Exports now finances over 90% import compared to only 60% in latter half of the 1980s.
- India's export was $310.5 billion in 2014-15. Although exports have fallen over the past one year in the wake of slowing global demand.
- Current account deficit was 3% of Gross Domestic Product (GDP) in 1990-91. It was less than 0.5% in 1994-95. Although, now it is 1.7% of GDP in 2014-15.
- International confidence in India has been restored.
- The share of industry in total employment increased from 16.2% to 22% in 2009.
- On the other side, the average annual growth rate of industrial production was 7.8% before globalization which fell to 6% during 1999-2000.

Index of Industrial Production (IIP) shows fluctuating trends. Growth reached 15.5% in 2007-08 and then started decelerating. Initial slowdown in the industrial growth was largely on account of global economic meltdown. There was, however, recovery in industrial growth from 2.5% in 2008-09 to 5.3% in 2009-10 and 8.2% in 2010-11.

During the last 2-3 years, the Indian industry has seen a rough patch decelerating considerably. The industrial growth fell from 9.2% (10th Five Year plan) to 7.2% (11th Five Year plan) and again to 0.35% (1st two year of 12th Five Year plan, FY 13 & 14).

However, with the new government in power, the industrial growth picked up from 0.1% in FY 14 to the average of 2.7% during the first half of 2014-15. Furthermore, the announcement made in Union budget 2014-15 are inspiring, as focus on industrial infrastructure and plan to establish 100 SMART cities would enhance industrialization and create employment opportunities in the economy. The launch of "Make in India" is expected to inspire the investors to look at India as their future investment destination, which would improve the "ease of doing business" and spur manufacturing growth in the coming years.

4.3 Causes of Unsatisfactory Industrial Performance

- Industrial sector which was almost totally protected was suddenly exposed to foreign competition through significant liberalization of import and drastic reduction in import duties. The industry was hardly prepared for it and slowdown was only to be expected.
- Consequently upon the adoption of the macroeconomic adjustment program of the International Monetary Fund (IMF) in 1991, the government of India was forced to cut down the public expenditure drastically. This led to depressing effect on private investment.
- Industrial production suffered on the account of inadequate availability of infrastructure like power, transportation and poor road conditions.
- Due to increasing competition in the international market, there is slow growth in exports. There is acute contraction in consumer demand. First, rural purchasing was severely affected by lower agricultural growth. Second, substantial wealth erosion caused by the fall in equities and real estate market also hampered the average urban consumer's proclivity to spend.

5. SUMMARY & CONCLUSION

The process of globalization in India has led to an unequal competition – a competition between giant MNCs and dwarf Indian enterprises. Thus, Indian entrepreneur suffers from sure disadvantage. Apart from that, cost of capital for Indian businesses is much higher than MNCs due to higher rate of interest in India. Restructuring and downsizing of Indian companies is not easy as labor laws doesn't allow easy retrenchment of labor. As against this, MNCs start new enterprises with modern technology and reduced requirement of labor. In some areas like taxes, the state has pursued policies that are clearly discriminating in favor of MNCs.

The overall analysis of globalization process discovers that the privatization may make for improved efficiency of public sector. A number of changes in industrial licensing policy, foreign investment, foreign technology agreement and MRTP Act are such as to do away with prior clearance of government. In such cases, project time and therefore, cost will be reduced and efficiency will improve. The policy of government of India should be able to bring foreign direct investment into manufacturing sector and high technology areas through which Indian economy can complete the globalization process.

Thus it can be concluded that flowing with globalization, India is shining in nearly every prospect. India is getting global recognition and slowly moving towards to become a major economic and political power.

REFERENCES

1. Ray, A., 1993. External Sector liberalization in India, Economic and Political Weekly, 2, 21-61.
2. Aggarwal, A.N., and Aggarwal, M.K., Indian Economy- Problem of Development and Planning, New Age International Publishers, New Delhi.
3. Nayyar, B.R., 2001. Globalization and Nationalism, 164.
4. Jalan, B., 1991. India's Economic Crisis: The way ahead out. New Delhi. 2-12.

5. Nayyar, D., 1988. Globalization: What does it mean for development? B. Debroy (Ed) Challenges of Globalization, New Delhi. 15.
6. Handbook of Industrial Policy and Statistics, Government of India, 2001.10.
7. Sandesra, J.C., 1991. New Industrial Policy: Question of efficient growth and several objectives, Economic and Political Weekly, p 187.
8. Economic Reforms: Two years after and a task ahead, discussion paper, Department of Economic Affairs, Ministry of Finance, Government of India, New Delhi, 1993, p 6.
9. Indian Economy-2015, Pratiyogita Darpan, Upkar Prakashan, Agra-2 (U.P.)
10. Various Economic Surveys of Government of India.
11. Puri, V.K., and Mishra, S.K., Indian Economy, Himalaya Publishing House Pvt. Ltd., Mumbai.

ISBN-13: 978-1533168078
ISBN-10: 1533168075

Chemistry of Industrial Globalization, Environmental Pollution and its Chem-Biological Significance

February 2016

Industrial Globalization and Environmental Awareness
Proceedings of the NATIONAL SEMINAR held by
Department of Chemistry, Government PG College, Ambala Cantt - Haryana - India

ACTIVATED CARBON ADSORPTION TECHNOLOGY FOR VOC EMISSION CONTROL

Rashmi Dhawan*[1], Sandeep Kumar[2], Anshul Bansal[3]

[1,3] *Department of Chemistry, S.A. Jain College, Ambala City*
[2] *Department of Chemistry, Dyal Singh College, Karnal*
*rashu.v3@gmail.com

ABSTRACT

Increasing industrialization, uncontrolled use and exploitation of natural resources during the last several decades has caused a major devastation of the earth, destruction of biodiversity and degradation of human health. These human activities have put a considerable pressure on the availability of basic human necessities like clean water and clean air. Huge amounts of obnoxious gases and vapors are being released into the atmosphere by industrial effluents and automobiles which is a major threat to man and his environment. As the variety and the amount of these chemical is ever increasing, the conventional methods of treatment become inefficient and sometimes ineffective. Consequently, there is need to develop alternative technologies for removal of inorganic and organic pollutants from air. Several laboratory experiments and field operations have shown that activated carbon adsorption is one such broad spectrum technology which has a great potential for the treatment of VOC and environmentally harmful gases before discharging into the air or recycling. This paper highlights the importance of activated carbon adsorption technology for VOC emission control. This article also presents a brief review on researches on adsorptive removal of volatile organic compounds and correlation of adsorption capacity with surface area & surface functional groups associated with different types of activated carbons.

Introduction

Rapid development of chemical, pharmaceutical and other process industries and the ever increasing number of automobiles on the roads together with products of the combustion of fossil fuels in power plants, chemical reactions and deliberate emission of chemical warfare agents is releasing large amounts of harmful gases and vapors into the air. The common hazardous air pollutants are flue gases such as NO_x, SO_x, CO, H_2S, NH_3, Hg vapors and other volatile organic compounds such as aromatic hydrocarbons, sulfur containing compounds, alcohol, ethers, hydrocarbons *etc.* A brief list of some common VOCs along with their physicochemical properties and permissible limit is presented in Table 1.

Table 1: Physicochemical Properties & threshold limit for exposure of some common VOCs

Name of Pollutant	Boiling Point (K)	Vapor Pressure (hPa) at 20 °C	Threshold limit of exposure (TLV) [1] (ppm)
Dimethyl sulfide	310.5	502	10
Benzene	353.1	101	0.5
Ethanol	351.5	59	100
Methanol	337.6	128	200
Toluene	483.3	22	50
p- xylene	411.3	8.7	100
Carbon disulfide	319.3	400	10
Dichloromethane	313.0	475	50
Chloroform	334.2	210	10
Carbon tetrachloride	349.7	120	5
1-propanol	370.0	21	100
1-butanol	390.6	6.3	20
Methyl Acetate	329.9	220	200
Ethyl acetate	350	97	400
Acetone	329.2	240	500
Diethyl ether	307.6	587	400
Methyl *t*-butyl ether (MTBE)	328.2	250	50
Hexane	342	160	50
Pyridine	388.5	20	5

[1]Extracted from Christian Reichardt, *Solvents and Solvent Effects in Organic Chemistry*, Wiley-VCH   Publishers, 3rd ed., **2003**, pages 501-502.

The VOC combine with sulphur and nitrogen oxides and other airborne chemicals in presence of light to produce photochemical smog and sometimes also produce acid rain/smog clouds. These emissions put considerable pressure on the availability of clean air and consequently, produce adverse environmental effects. People exposed to larger concentrations of these hazardous air pollutants may experience serious health effects which include severe skin and eye infection, pulmonary edema, neurotoxicity, carcinogenesis, teratogenesis and mutagenesis or potentially fatal condition [1-3]. It is estimated that about three million deaths occur globally each year due to air pollution, mainly by particulate matter (organic & inorganic). To comply with stringent air regulations, efforts are being made to develop newer technologies for removal of VOC from the effluent. Industrial pollution control solutions involve ventilation, collection/recovery, control and destruction of the hazardous volatile organic compounds. But these conventional methods have their associated disadvantages. Among these processes, adsorption has been found to be a feasible and an effective method to separate and recover VOCs for further analysis. Consequently, a considerable amount of work has been carried out for removal of these air pollutants by adsorption technology. The control of VOC emissions typically reduces the concentrations from between 400 and 2,000 parts per million (ppm) to under 50 ppm. Adsorption technology can now extend the range of VOC concentration from 20 ppm to one-fourth of the Lower Explosive Limit (LEL). At the lower end of this range, such small concentrations may be difficult or uneconomical to control by another technology In literature, several adsorbents such as zeolites, silica, micro porous clay and carbon materials are reported for the adsorptive removal of these vapors [4-8]. Among these adsorbents, activated carbons are excellent adsorbents due to high surface area, highly developed micro porosity and their tunable surface functional groups. Activated carbon is promising material for treatment of a wide variety of hazardous environmental contaminants from air before atmospheric emission.

Application of activated carbon for adsorptive removal of volatile organic compounds

Volatile Organic Compounds (VOCs) are among the most common air pollutants emitted from chemical, petrochemical and allied industries. The presence of even small amount of these compounds cause environmental hazards and risk to plant, animal and human health. Activated carbons have been extensively investigated as adsorbents for removal of VOC from airstream by different research groups.

Singh et.al [9] studied the adsorption behavior of hexane and benzene in single component and mixture system on an activated carbon cloth and observed that internal diffusion controlled the adsorption of hexane and benzene. Dastgheib and Foster et al. [10] examined the adsorption characteristics of ACF with different specific surface areas towards VOC (acetone, benzene, n-butane) over a range of concentrations in air streams. The results showed that at lower concentrations, the ACF with the least specific surface area adsorbed more but at saturation, adsorption increased with increasing specific surface area due to the widening of micropore. Xie et al. [11] measured the isotherms of toluene, butanol, ethylacetate on monolith activated carbon at different temperatures using a microbalance. The experimental data was correlated by the Langmuir, Freundlich, Dubinin-Raduhkevich and Toth adsorption equations. The results showed that monolith activated carbon had a high adsorption capacity for these three VOCs. Park et al. [12] measured the adsorption equilibrium of toluene, dichloromethane and trichloroethane on pitch based ACF at different temperatures and found that among several adsorption isotherm models, Toth equation agreed remarkably with the experimental data. Isosteric heats of adsorption indicated that activated carbon had an energetically heterogeneous surface. Qi-bin et al. [13] discussed the influence of metal ions doped onto adsorption of dichloromethane and trichloromethane on activated carbons with the help of the hard soft acid base principle. The results showed that the adsorption decreased in the order of Fe(III)-AC > Mg(II)-AC > Cu(II)-AC > AC > Ag(I)-AC. Cuervo et al. [14] studied the influence of oxygen functional groups on the adsorption of different alkanes, aromatics and chlorohydrocarbons on high surface area graphite and carbon nanofibre (CNF) after oxidation with nitric acid and heat treatment at 900°C. It was found that the adsorption capacity decreased after the oxidative treatment of the graphites. These workers observed that the presence of oxygen surface groups influenced the adsorption of aromatics and double-bonded compounds while adsorption of n-alkanes and cyclic compounds was not influenced at all. Cui et al. [15] evaluated the ability of activated carbons for the adsorptive removal of other sulfur compounds such as H_2S, methyl mercaptan, ethyl mercaptan, dimethyl disulfide, diethyl disulfide and tetrahydrothiophene after modification of activated carbons by oxidation and impregnation methods and found that the adsorption increased after each modification. Guo et al. [16] studied the adsorption of CS_2 on one raw and modified activated carbon sample at different temperatures from 303 K to 313 K and found that the amount of carbon disulfide adsorbed by the modified carbon samples was larger than that of raw one. Zhou et al. [17] examined the effects of modification of activated carbons (ACs) by HNO_3 oxidation and gas-phase O_2 oxidation on the adsorption of sulfur compounds such as benzothiophene, dibenzothiophene, 6-methyldibenzothiophene and 4-methylbenzothiophene. The improved adsorption performance on oxidation with HNO_3 was attributed mainly to an increase in the acidic oxygen-containing functional groups. The correlation between the concentration of the surface oxygen functional groups and the adsorption capacity per unit area suggested that the adsorption involved an interaction of the acidic oxygen-containing groups on activated carbons with the sulfur compounds. Chen et al. [18] studied the adsorption behavior of pyridine vapor on coals and WK-11 ion exchange resin. The results indicated that pyridine could disrupt weak hydrogen bonds of original WK-11 to form strong pyridine-COOH hydrogen bond. Gunko and Bandosz [19] measured the adsorption isotherms of water and methanol volumetrically and of diethylether by inverse gas chromatography. The results showed that the effect of surface chemistry and the presence of oxygenated groups are predominant in the case of water vapor adsorption and of least important in the case of adsorption of diethylether. Chen and Wang [20] studied the correlation between physicochemical properties of volatile organic compounds (toluene, p-xylene, methylacetate, ethylacetate, ethanol and propanol) and saturation adsorption capacity of activated carbon. The adsorption isotherms were fitted with Langmuir equation. The saturation adsorption capacity was maximum for toluene and minimum for methyl acetate. Yi et al. [21] determined the adsorption of several volatile organic compounds on ACF modified by $CuSO_4$. The results indicated that modified ACF showed higher adsorption capacities for benzene, toluene, methanol and ethanol than the untreated ACF. Yu et al. [22] studied the adsorption of MTBE and other fuel oxygenates on two bituminous coal activated carbons and found that adsorption capacity of MTBE was lower as compared to other oygenates.

Conclusions

In summary, Activated carbons have a great potential for the removal of vapors from working places of various industries and chemical laboratories due to their high adsorption capacity. Furthermore, due to its tunable porous and chemical structure, its surface can be modified according to its application towards particular group of pollutants. However, the study of removal mechanism and the effect of various parameters are necessary to optimize the adsorptive capacity of activated carbon. Activated carbon adsorption technology is a proven remediation technology that is simple to install, easy to operate and maintain and have applications in maintaining and improving ambient air quality.

References:
1. Leikauf G.D., *Environ Health Perspect.* 110, 505–526, (2002).
2. Buka I., Koranteng S., Osomio-Vargas A.R., *Paediatr Child Health.*, 11, 513–516, (2006)
3. Ozkayanak H., Palma T., Touma J.S., Thurman J., *J. Expo. Sci. Environ. Epidemiol.* 18, 45–58, (2008).

4. Hernandez M.A., Corona L., Gonzalez A.I., *Ind. Eng. Chem. Res.,* 44, 2908-2916, (2005).
5. Diaz E., Ordonez S., Vega A., Coca J., *J. Chromatogr. A*, 1049, 139-146, (2004).
6. Chihara K., Terakado T., Ninomiya T., Mizochi H., *Stud. Surf. Sci. Catal.*, 154 B, 1991-1998, (2004).
7. Ulendeeva A.D., Lygin V.I., Lyapina N.K., *Kinet. Katal.*, 20, 978-983, (1979).
8. Khattak A.K., Mahmood K., Afzal M., Saleem M., Qadeer R., *Colloids and Surfaces A: Physicochem. Eng. Aspects*, 236, 103-110, (2004).
9. Singh K.P., Mohan D., Tandon G.S., Gupta G.S.D., *Ind. Eng. Chem. Res.*, 41, 2480-2486, (2002).
10. Foster K.L., Fuerman R.G., Economy J., Larson S.M., Rood M.J., *Chem. Mater.,* 4, 1068-1073, (1992).
11. Xie L., Luo L., Li Z., *Huagong Xuebao*, 57, 1357-1363, (2006).
12. Park J.W., Lee S.S., Choi D.K., Lee Y.W., Kim Y.M., *J. Chem. Eng. Data*, 47, 980-983, (2002).
13. Qi-bin X., Si-si H., Li-ming X., Zhang W., Li Z., *Gongneng Cailiao*, 40, 1911-1914, (2009).
14. Cuervo M.R., Asedegbega-Nieto E., Diaz E., Vega A., Ordonez S., Castillejos-Lopez E., Rodriguez-Ramos I., *J. Chromatogr. A,* 1188, 264-273, (2008).
15. Cui H., Reese M.A., Turn S.Q., *Prepr. Symp. Am. Chem. Soc., Div. Fuel Chem.*, 52, 680-681, (2007).
16. Guo B., Chang L., Xie K., *Fuel Process. Technol.*, 87, 873-881, (2006).
17. Zhou A., Ma X., Song C., *Appl. Catal. B,* 87, 190-199, (2009).
18. Chen H.K, Miura K., Li W., Li B., *Ranliao Huaxue Xuebao*, 32, 135-139, (2004).
19. Gun'ko V., Bandosz T.J., *Phys. Chem. Chem. Phys.,* 5, 2096-2103, (2003).
20. Chen L., Wang J., *Huagong Huanbao*, 27, 409-412, (2007).
21. Yi F.Y., Xiao-Dan L., Shui-Xia C., Xiao-Qun W., *J. Porous Mater.*, 16, 521-526, (2009).
22. Yu L., Adam C., Ludlow D., *J. Environ. Eng.*, 131, 983-987, (2005).

ISBN-13: 978-1533168078
ISBN-10: 1533168075

Chemistry of Industrial Globalization, Environmental Pollution and its Chem-Biological Significance

February 2016

Industrial Globalization and Environmental Awareness

Proceedings of the NATIONAL SEMINAR held by
Department of Chemistry, Government PG College, Ambala Cantt - Haryana - India

Green Approach In The Synthesis Of Heterocyclic Compounds From α-Tosyloxyketones and α,β-Ditosyloxyketones

Dr Deepak Sharma

Rajiv Gandhi Govt. College, Saha, Ambala
drdeepak2108@rediffmail.com

Abstract

α-Tosyloxyketones and α,β-ditosyloxyketones are important precursors thereby providing green and superior alternative to large number of heterocyclic syntheses that make the use of highly lachrymatory bromo analogs. The present paper is a review article that emphasizes the synthesis of heterocycles which involve little or no toxic chemical substances. This review article highlights the synthesis of five membered heterocycles from monotosyloxy and ditosyloxycarbonyl compounds which are less toxic as compared to their bromo analogs. The approach of using monotosyloxy and ditosyloxyketones as intermediate has distinct advantage that it not only avoids the use of highly toxic α –haloketones, but also provides a direct one step synthesis of various heterocycles.

Keywords: α-Tosyloxyketones, α,β-ditosyloxyketones, HTIB and Heterocyclic Compounds.

The concept of Green Chemistry incorporates a new approach[1-6] to the synthesis, processing and application of chemical substances in such a manner as to reduce threats to human health and environment. Since one of the twelve principles of green chemistry is that the synthetic methods should be designed to use and generate substances that possess little or no toxicity to human health and the environment, the present paper is a review article that emphasizes on the alternative routes to the conventional one for the synthesis of heterocycles which involve little or no toxic chemical substances. This review article highlights the synthesis of five membered heterocycles from monotosyloxy and ditosyloxycarbonyl compounds which are less toxic as compared to their bromo analogoues.

α-Tosyloxycarbonyl compounds are important precursors for the synthesis of a number of heterocyclics. Infect the studies involving α-Tosyloxyketones have offered a superior alternative to the large number of existing syntheses that make the use of highly lachrymatory α -haloketones in the conventional approach. The approach of using α-tosyloxyketones as intermediate has distinct advantage that it not only avoids the use of highly toxic α –haloketones, but also provides a direct one step synthesis of various heterocycles.

α –Aryloxyketones **3**, obtained by the reaction of α –tosyloxyacetophenones with phenols, undergo cyclization to benzofurans (**4**) by using the standard conditions (**Scheme-1**).[7]

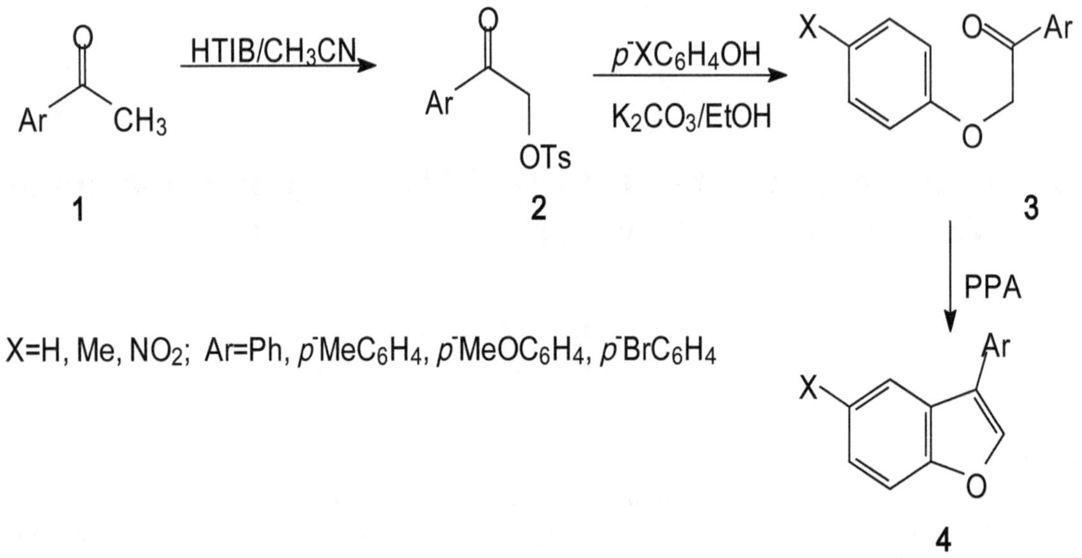

X=H, Me, NO$_2$; Ar=Ph, pMeC$_6$H$_4$, pMeOC$_6$H$_4$, pBrC$_6$H$_4$

Scheme-1

2-(α-Tosyloxy)acylphenylbenzoates (**6**) are cyclized to coumarin-3-onedimethylacetals (**7**) by using KOH in methanol. Acid hydrolysis of **7** affords the corresponding coumarin-3-ones (**8**) (**Scheme-2**).[8]

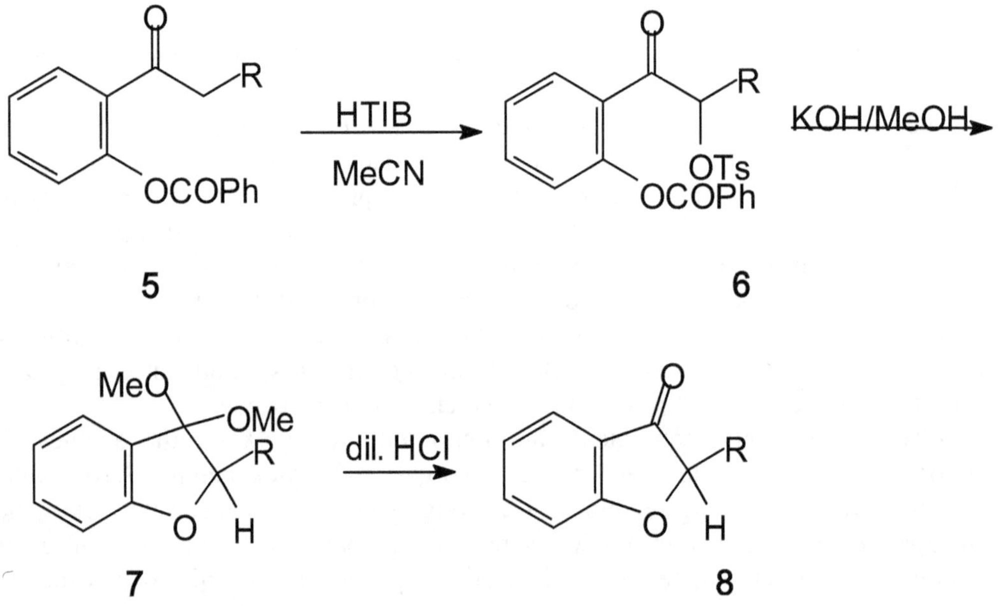

Scheme-2

Pyrazoles **11** are synthesized by the oxidation of arylhydrazones with HTIB. The reaction was carried out in one pot in the presence of diisopropylethylamine. In the intermediate α-tosyloxycompounds (generated in situ), the intramolecular participation of amino group displaces the tosyloxy group to produce the cyclized products (**Scheme-3**).[9]

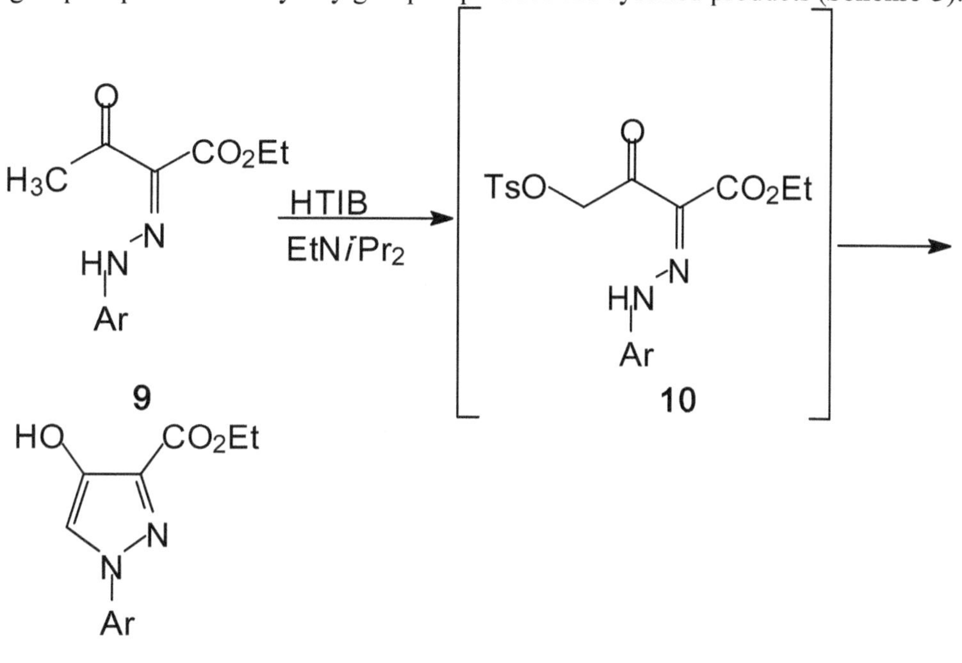

Scheme-3

α-Aryloxyacetophenones (**12**), which are obtained by the oxidation of **1** with HTIB by treatment with *p*-substituted benzoic acids, are cyclized to oxazoles (**13**) (**Scheme-4**).[7]

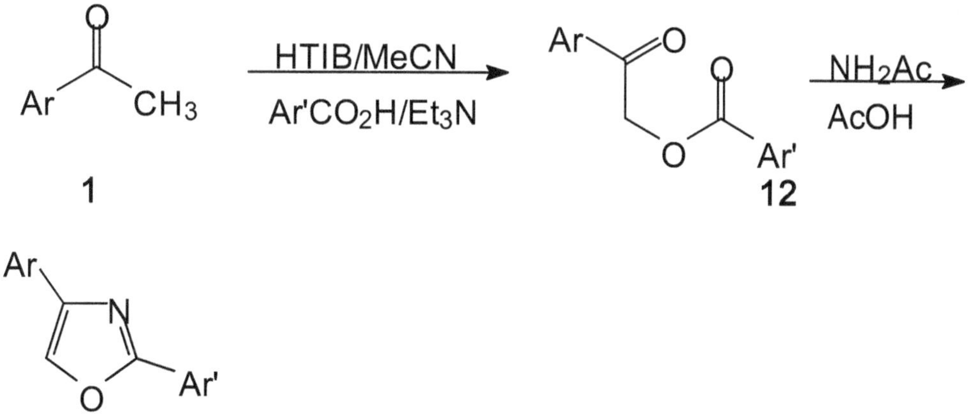

Scheme-4

HTIB mediated method has been successfully used for the synthesis of 2-mercaptoimidazoles. The method offers a superior alternative to the most widely used Marckwald's synthesis.[10] The method is also applicable to the synthesis of 4-(2-thienyl)imidazoles(**14**, R=2-thienyl) (**Scheme-5**).[11]

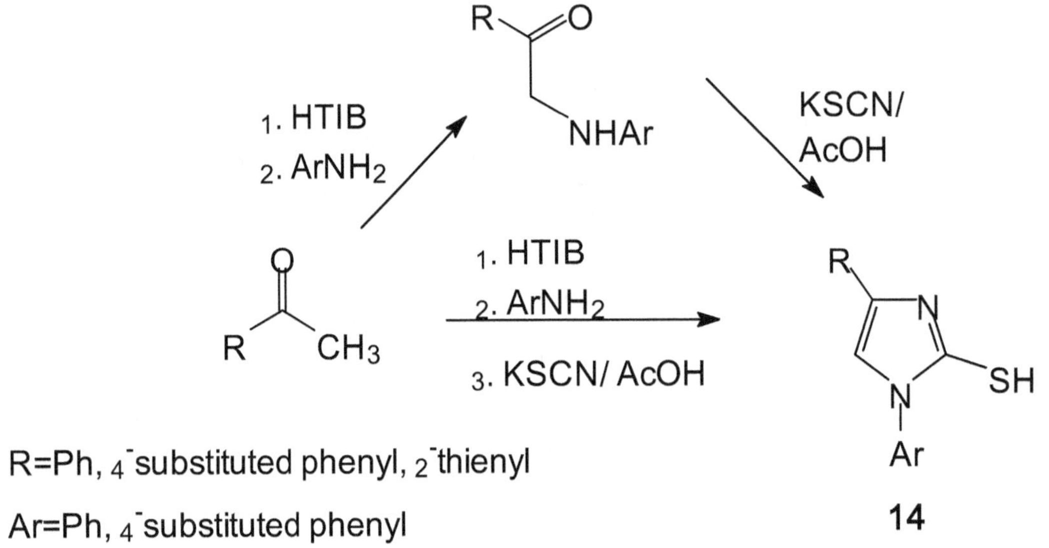

R=Ph, 4⁻substituted phenyl, 2⁻thienyl

Ar=Ph, 4⁻substituted phenyl

Scheme-5

The condensation of α-tosyloxyketones **15** with thioureas (substituted thioureas) and thiamides produces 2-amino (substituted amino) (**16**) and 2-alkyl/arylthiazoles (**17**) respectively.[12-15] These syntheses provide a superior alternative of the well known Hantzsch thiazole synthesis.[16,17] The one pot procedure are also employed for these syntheses starting from corresponding ketones (**Scheme-6**).

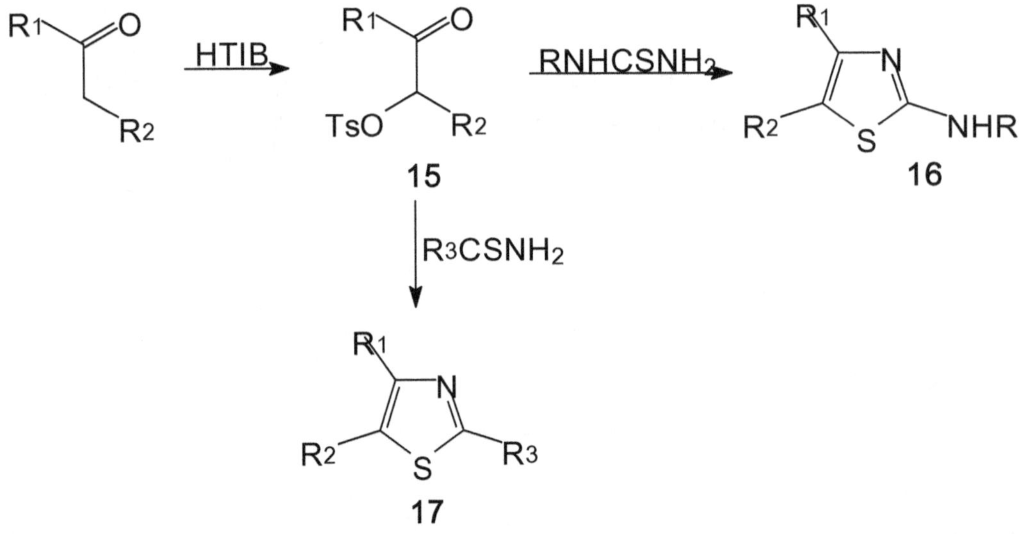

Scheme-6

Further, α,β-ditosyloxyketones **18** provide another green alternative to the conventional one for the synthesis of heterocyclics. The reaction of α,β-chalcone ditosylates **18** with various reagents such as phenylhydrazine hydrochloride, semicarbazide hydrochloride and thiosemicarbazide in suitable conditions leads to 1,2-aryl shift, thereby providing a novel route for the synthesis of 1,4,5-trisubstituted pyrazoles[18] **19** whereas the reaction with hydroxylamine hydrochloride provides a conveniont route to 4,5-disubstituted isoxazoles **20**(**Scheme-7**).[19]

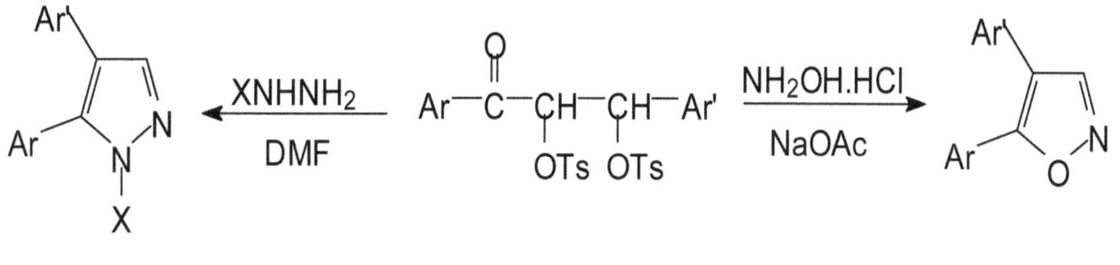

19   **18**

X=Ph, $CONH_2$, $CSNH_2$

Scheme-7

Conclusion

α-Tosyloxyketones and α,β-ditosyloxyketones are green alternative to their highly lachrymatory bomo analogs. The green route involves simple experimentation and yields are fairly good.

References

1. Anastas, P.T.; Hovarsth, I.T., *Innovations and Green Chemistry, Chem. Rev.* 2007, **107**, 2169.
2. Ravichandran, S., *Int. J. Chem. Tech. Res.*, 2010, **2(4)**, 2191.
3. Sheldon, R.A., *Green Solvents for sustainable Organic Synthesis: State of the art. Green Chem.*, 2005, **7**, 267.
4. Bharati, V.B., **Resonance**, 2008, 1041.
5. Ahluwalia, V.K.; Kidwai, M., *New trends in Green Chemistry, Anamaya Publisher, New Delhi*, 2004.
6. Dhage, S.D., *IJRPC*, 2013, **3(3)**, 518.
7. Parkash, O.; Saini, N.; Sharma, P.K., *J. Indian Chem. Soc.*, 1995, **72**, 129.
8. Moriarty, R.M.; Parkash, O. *Advances in Heterocyclic Chemistry*, 1998, 69, Chapter 1, pp 1-84.
9. Patel, H.V.; Vyas, K.A.; Padey, S.P.; Tavares, F.; Fernandes, P.S., *Synth. Commun.*, 1991, **21**, 1583.
10. Marckwald, *Chem. Ber.*, 1982, **25**, 2354.
11. Parkash, O.; Aggarwal, R.; Saini, N.; Goyal, S.; Tomer, R.K.; Singh, S.P. *Indian J. Chem.* 1994, **33B**, 116.

12. Parkash, O.; Goyal, S. *Indian J, Heterocyclic Chem.*, 1991, **1**, 99.
13. Parkash, O.; Aggarwal, R.; Singh, S.P., *Indian J, Heterocyclic Chem.*, 1992, **2**, 111.
14. Moriarty, R.M.; Vaid, B.K.; Duncan, M.P.; Devi, S.O.; Parkash, O.; Goyal, S., *Synthesis*, 1992, 845.
15. Parkash, O.; Saini, S.; Saini, N.; Prakash, I.; Singh, S.P., *Indian J. Chem.*, 1995, **34B**, 660.
16. Hantzsch, A; Weber, H. *Chem. Ber.*, 1887, **20**, 3118.
17. Hantzsch, A; Weber, H. *Chem. Ber.*, 1888, **21**, 938.
18. Parkash, O.; Sharma, D.; Kamal, R.; Kumar, R.; Nair, R.R., *Tetrahedron*, 2009, **65**, 10175.
19. Kamal, R.; Sharma, D.; Wadhwa, D.; Parkash, O., *Synlett*, 2012, **23**, 93.

ISBN-13: 978-1533168078
ISBN-10: 1533168075

Chemistry of Industrial Globalization, Environmental Pollution and its Chem-Biological Significance

February 2016

Industrial Globalization and Environmental Awareness

Proceedings of the NATIONAL SEMINAR held by
Department of Chemistry, Government PG College, Ambala Cantt - Haryana - India

The Environmental Impact of E-waste

Dr Ramesh Sandhu

Associate Professor
C. R. College of Education, Hisar.

Abstract

Electronic waste or e-wasting is growing exponential around the world due to obsolescence and consumerism. E-waste is the fastest growing waste stream and most of which is dumped in the general waste heap. E-waste consists of both types; electrical and electronics appliances. Electrical appliances are fridges, A.Cs, washing machines, microwave ovens and fluorescent lights while electronics equipments comprises of computers and their accessories, mobile phones, TVs and stereos etc. The Organisation for Economic Co-operation and Development (OECD) defines e-waste as" any appliance using an electric power supply that has reached its end-of-life ". The composition of e-waste is very diverse and differs across product lines and categories. Overall, it contains more than 1000 different substances which fall into "hazardous, and "non-hazardous" categories; significantly, the toxicity of many of the chemicals in e-waste is unknown (Lundergren, 2012). There are several issues related to e-waste like high volumes, Toxic design, poor design and complexity, labour issues, financial incentives and lack of regulations. Presence of heavy metals, POPs, flame retardant and other potentially hazardous substances poses risks to human health and environment. Toxic substances are generated by emissions or output of leachates from dumping activities, particulate matter, fly and bottom ashes while burning, fumes from mercury amalgamate, effluents from cyanide leaching etc. It is very complex and difficult Task to recycle and manage e-waste. The effective e-waste management will depend upon the cooperative effort of local, national, regional and international initiatives.

Key words: E-waste, health hazards.

Introduction

Information and communication revolution has brought wide changes in our economies, society, industries and institutions. These innovative technologies have enhanced the quality of our lives but at the same time created many problems. The problem of e-waste has become an immediate and long term concern as its unregulated and improper accumulation and recycling can lead to major environmental problems endangering not only human and animal health but also environment health due to toxic and other dangerous materials available in them (Binay, 2012). E-waste consists of both electronic and electrical appliances. It comprises of computers, mobile phones, digital music recorders/players, refrigerators, washing machines, televisions and many other household consumer items. It includes used electronics that can be reused, resold, salvaged, recycled or disposed off. Others are re-usable (working and repairable electronics) and secondary scrap (copper, steel, plastic etc.) which is known as commodities. The residue or material that is discarded by the buyer rather than recycled is termed as waste. E-waste has been categorised into three main components i.e. Large Household Appliances (ii) IT & (iii) Telecom and consumer equipments. Refrigerator and washing machine represent large household appliances; PC, monitor and laptop represent IT and Telecom, while TV represents Consumer Equipments. Each of these e-waste items has been classified with respect to 26 common components found in them. These components form the building blocks' of each item and therefore they are readily 'identifiable' and 'removable' (Sitaramaiah et al. 2014). The popularity of consumer electronics is increasing day by day. The advent of smart phones, i-pads, e-readers, e-notes takers and other smart devices in daily and professional life have made these items a necessity. Thus, the rapidly growing e-waste is the main challenge. It is predicted that, by 2018, more PCs will be discarded in countries in the developing than the developed world, and the prevailing assumption that trade is the main driver of informal recycling might soon become obsolete (Bett, 2010). It is predicted that, by 2020, in both china and South Africa, there will be 200-400 percent more e-waste from old computers than in 2007, and a staggering 500 percent more in India. The volume of e-waste from discarded mobile phones will be about seven times higher in china and is 18 times higher in India (Lundgren, 2012). When adding the vast amounts of e-waste that are still being imported to these countries, both legally and illegally, it is evident that the problem is exploding, with many dangers for human health and the environment (McCarthy, 2010). There are several challenges associated with e-waste. They include the unwillingness of consumers to return and pay for disposal of used electronic products; the uncoordinated, high level of importation of e-waste disguised as second hand devices; a lack of awareness among consumers; a lack of awareness of the potential hazards of e-waste among collectors and recyclers; a lack of funds and investment to finance improvements in e-waste recycling; the absence of recycling infrastructure or appropriate management of e-waste; the absence of effective take-back programmes; the lack of interest by companies or incentives for e-waste management; and the failure of and/or lax implementation of e- waste-specific legislation (chi, 2011). Cyber crime is another challenge posed by information and communication technologies.

Health hazards due to chemicals in e-waste.

Toxic substances reside in circuit boards, mother boards, solders, batteries and LCDs. Some dangerous substances and their concentration are mentioned below.

Arsenic is found in semi conductors, diodes, microwaves, LEDs, & solar cells. It causes arsenic poisoning. Effect of arsenic can take years to develop, include skin lesions, peripheral neuropathy, gastro-intestinal symptoms, diabetes, renal system effects, cardiovascular diseases and cancer (WHO, 2010b).

Barium is found in electron tubes, filler for plastic and rubber and lubricant activities. Short-term exposure causes muscle weaknesses and damage to heart, liver and spleen. It also produces brain swelling after short-exposure (Osuagwu & Ikerionwu, 2010).

Cadmium is used in batteries, pigments, solder, alloys, circuit boards, computer batteries & monitor cathode ray tubes (CRTs). Cadmium has toxic, irreversible effects on human health and accumulates in kidney and liver (op. cit.). It has toxic effects on kidney, skeletal system and the respiratory system, and is classified as a human carcinogen (WHO, 2010c).

Cobalt is found in rechargeable batteries and coatings for hard disk drives. It is hazardous in case of inhalation and ingestion, and is an irritant of the skin. It has carcinogenic effects and is toxic to lungs. Repeated on prolonged exposure can produce target organ damage (Material safety Data Sheet, 2005).

Dioxins are produced when electronics are burnt in open air. It is highly toxic and can cause chloracne, reproductive and developmental problems, damage the immune system, interfere with hormones and cause cancer (WHO,2010 d).

Hexavalent Chromium is used as corrosion protection of untreated and galvanised steel plates and a decorator on hardner for steel housings (Osuagwu & Ikerionwu, 2010). It damages kidneys, the liver and DNA. Asthmatic bronchitis has been linked to this substance (Osuagwu & Ikerionwu, 2010). It causes asthma and damages the skin, liver and kidney. It may increase or reduce blood leukocytes, eosinophilla and causes lung cancer.

Lead is found in solder of printed circuit boards, glass panels and gaskets in computer monitors (Osuagwu & Ikerionwu, 2010). It causes damage to central and peripheral nervous systems, blood systems and kidneys, and affects the brain development of children. (Osuagwu & Ikerionwu, 2010). A cumulative toxicant that affects multiple body systems, including the neurological, haematological, gastrointestinal, cardiovascular and renal systems (WHO, 2010 e).

Lithium is used in rechargeable batteries. Lithium ingestion is extremely harmful because it passes through the placenta, It is hazardous and an irritant of the skin and eye, and when inhaled, it can be excreted in maternal milk (Material safety Data sheet, 2005).

Mercury is found in relays, switches, CFL and printed circuit boards (Osuagwu & Ikerionwu, 2010). Elemental and methyl-mercury are harmful to central and peripheral nervous system. Mercury fumes are harmful to nervous, digestive and immune systems, lungs and kidneys and may be fatal. The inorganic salts of mercury are corrosive to the skin, eyes and gastrointestinal tract, and may induce kidney toxicity if ingested (WHO, 2007).

Polybrominated diphenyl ethers (PBDEs) are used in brominated flame redundant (BFRs) and found in plastic housing of electronic equipments and circuit boards to reduce flammability (Tsydenova & Bengtsson, 2011). PBDEs are considered dangerous because these substances have high lipophilicity and high resistance to degradation processes. Heptatoxicity, embryotoxicity and thyroid effects seem to be characteristic end points in animal toxicity, and behavioural effects have been demonstrated (Darnerud, Eriksson, Johannesson, Larsen and Vilekrela, 2001). BFRs in general have been shown to disrupt endocrine system functions and may have effect on the levels of thyroid stimulating hormone and cause genotoxic damage, causing high cancer risk (Tsydenova & Bengtsson, 2011).

Polyvinylchloride (PVC) is used in cabling and computer housing plastics due to its fire retardant qualities. PVC produces dioxins when burnt; causes reproductive and developmental problems, immune system damage and interferes with regulatory hormones (Osuagwu & Ikerionwu, 2010).

Polychlorinated biphenyls (PCBs) are used in insulating material in older electronic products. PCBs are linked to reproductive failure and suppression of the immune system (Stockholm Convention n-d.).

Silver is used in wiring circuit boards etc. and is very hazardous to eyes in case of eye contact, ingestion and inhalation. Severe over-exposure can result in death. Repeated exposure may produce general deterioration of health by an accumulation In one or many human organs (Material Safety Data Sheet. 2005).

Zinc (Chromates) are found in plating material. Zinc in contact with eyes can cause irritation; powdered zinc is highly flammable (University of Oxford, 2005); if inhaled, causes a cough, and if ingested, abnormal pain, diarrhoea and vomiting is common (ICSC database n-d). This is not an exhaustive list of pollutants. There are many more chemical & elements used in manufacturing electronic items and these substances are also harmful to our health and environment. It is not threatening health of workers and general public but will have bad effects on future generations also. E-waste therefore constitutes a significant global environment and health emergency, with implications far broader than occupational exposure and involving vulnerable groups and generations to come (Sitaramaiah et al. 2014).

Treatment and Disposal

E- waste is composed of valuable substances. Thus, specialised and "high tech" methods are required to process e-waste, so that, we may recover maximum resources and minimise the toxicity risks to humans and environment. However, these specialised and "high tech" methods are limited to developed nations and developing countries are using crude techniques to extract precious material. This creates localised pollution of environment. Following methods can be used for waste management.

Recycling of e-waste

Several equipments like washing machine, fridge, TVs, computers, laptops, Monitor & CRT, Photostat machine, modems, telephones, hard discs, mobile phones, fax machines, printers, CPU, memory chips, connecting wires and cables can be recycled.

Recycling is also concerned with the dismantling of important parts i.e. different components of e- waste comprising of dangerous substances like PCB, Hg, PVC, CRT, ferrous and non-ferrous and printed circuits boards are segregated. Precious metals like Copper, gold, silver and palladium are extracted through the use of strong acids. Advanced & high end technologies can help to recover more amounts of precious elements.

Re-use

It involves the direct second hand use after minor modifications in electronic gadgets like desk top computers, cell phones etc., inkjet cartridge is also used for refilling. Old computers can be upgraded or donated to schools in rural areas, computers that cannot be repaired should be returned back to manufacturers. It will diminish the volume of e-waste generation. Buy back policy of old electronic equipments should be made mandatory. This policy will considerably reduce the volume of e-waste generation.

Reduce

E-waste can be reduced by Smart procurement and good maintenance.

Incineration

Although this method is not free from risks of pollution because the waste material is burnt in control and complete combustion process in specialty designed incinerators at a high temperature (900-1000^{0}C). However, it reduces the waste volume and utilise the energy content of combustible materials. This method converts the more environmentally hazardous Organic substance into less hazardous compounds.

However, some opportunities are also associated with e-waste. Recovery of rare earth elements from e-waste could boost e-waste recycling in the future. The cost of recycling is decreasing in developed countries. Other opportunities are in green design, innovations, life-cycle analysis, public outreach, social policy and so on. Some companies are already starting to design product content for the reality of waste handling operations in developing countries, for instance, by leaving out chemicals that can create hazardous pollutants when burned inappropriately (Nimpuno & Scruggs, 2011).

Reference

1. Betts, K. (2010). A changing e-waste equation, *in Environmental science and Technology*, Vol. 44, No. 9, p. 3204.
2. Binary Kumar (n.d.) e-waste-Environment and Human Health Hazards and Management. *IR SEE*/Prof.(Network Management)/NAIR,Vadodara.
3. Chi,X. (2011), Informal electronic waste recycling: A sector review with specific special focus on china, *in Waste Management*. Vol. 31, No. 4, p. 731-742.
4. Darnerud, P.; Friksen, G.; Johannesson, T.; Larsen, P.; Vileksela, M. (2001), Polybrominated biphenyl ethers: Occurrence, dietary exposure, and toxicology, *in Environmental Health Perspectives.* Vol. 109, supplement 1: Reviews in Environmental Health, Mar., pp. 49-68.
5. ICSC database, n.d. Available: http://www.ilo.org/dynicsc/showcard.home
6. Lunder,K.(2012). *The global impact of e-waste: Addressing the challenge*. Geneva, Softwork and SECTOR, ILO.
7. Material Safety Data Sheet (2005). *Material Safety Data Sheet Listing*. Available: http:// www. sciencelab.com/msdsList.php.
8. McCarthy, M. (2010).The Big Question: How big is the problem of electronic waste, and can it be tackled; in *The Independent*, 24 Feb, Available: http://www.independent.co.uk/environment/green-living/the-big-question-how-big-is-the-problem-of-electronic-waste-and-can-it-be-tackled-1908335.html.
9. Nimpuno, M.; Scruggs, C. (2011). *Information on chemicals in electronic products: A study in needs, gaps, obstacles and solutions to provide and access information on chemicals in electronic products* (TemaNord). Available:http://www.chemsec.org/get-informed/global-initiative/unep-cip-study-on-electronics.
10. Organisation for Economic Co-operation and Development (OECD) n. d. *Perfluorooctane sulfonate (PFOS) and related chemical products.*Available:http://www.oecd.org/document/58/0,3746,en_2649_34375_2384 378_1_1_1_11,00.html.

11. Osuogwu, 0. E.; Ikerionwu, C. (2010). E-cycling e-waste: The way forward for Nigeria IT and electro-mechanical industry *in International Journal of Academic Research.* Vol. 2, No. I, pp. 142-149.

12. Sitaramaiah, Y. Kusuma Kumari M. (2014), Impact of electronic waste leading to environmental health Pollution. *Journal of Chemical and Pharmaceutical Sciences. Special Issue 3.*

13. Stockholm convention. n.d. *About the convention,* Available: http://chm.pops.int/convention/tabid/54/Default.aspx.

14. Tsydenova, 0; Bengtsson, M. (2011). Chemical hazards associated with treatment of waste electrical and electronic equipment, *in Waste Management,* Vol. 31, No, I, pp. 45-58. University of Northampton, n.d. Electronic waste problems, Available://www. Northampton.ac.uk/download/3014/electronic-waste-problem.

15. World Health Organisation (WHO) (2007). *Exposure to mercury: A major public health concern.* Available: http://www.who.int/ipcs/features/mercury.pdf

16. World Health Organisation (WHO) (2010 b). *Exposure to arsenic: A major public health concern.* Available: http://www.who.int/ipcs/features/arsenic.pdf.

17. World Health Organisation (WHO) (2010 c). *Exposure to cadmium: A major public health concern.* Available: http://www.who.int/ipcs/features/cadmium.pdf.

18. World Health Organisation (WHO) (2010 d). *Exposure to dioxine and dioxine-like substance: A major public health concern.* Available: http://www.who.int/ipcs/features/dioxine.pdf.

19. World Health Organisation (WHO) (2010 e). *Exposure to leaf: A major public health concern.* Available: http://www.who.int/ipcs/features/lead.pdf.

ISBN-13: 978-1533168078
ISBN-10: 1533168075

Chemistry of Industrial Globalization, Environmental Pollution and its Chem-Biological Significance

February 2016

Industrial Globalization and Environmental Awareness

Proceedings of the NATIONAL SEMINAR held by

Department of Chemistry, Government PG College, Ambala Cantt - Haryana - India

Oxidative Coupling of Tetrahydroisoquinoline with Alkynes

Yogita Maheshwary

D. A. V. College for Girls, Yamunanagar

Abstract

Oxidative coupling of N-methyltetrahydroisoquinoline and alkynes using CuI–DEAD is studied. It gave the regioselective C-1- alkynylated products in good yield. This regioselectivity is opposite to the results obtained in the reaction of N,N-dimethylbenzyl amine where the N-methyl alkynylated product was formed exclusively or mainly. The C-1-substituted propargylic isoquinolines were smoothly reduced to phenethylisoquinolines with Pd/C. This reaction series gives a short route to synthesis of methopholine, homolaudanosine and other phenethylisoquinoline alkaloids.

Introduction

In many natural products and medicinally active compounds, N-methyltetrahydroisoquinoline represents the main moiety.[1] The C-1-functionalized derivatives of this molecule, both naturally occurring and synthesized ones, have been highly important in biological study as well as in medicinal developments.[2] 1- Alkynyl N-methyltetrahydroisoquinoline compounds are potential D3 dopamine receptor ligands in neurological and neuropsychiatric therapeutics.[3] Phenethylisoquinoline alkaloids,[4] a class of homologated 1-benzyl isoquinoline alkaloids, have also evinced considerable interest.[5] An efficient method for the synthesis of these alkaloids is, therefore, highly attractive.

Literature Search

Transition-metal-catalyzed cross-coupling reactions have been widely studied[6] and CDC (cross dehydrogenative coupling) technique for activation of C–H bond (sp, sp^2 , sp^3) is an effective method for construction of new C–C bonds.[6,7] Chao-Jun Li and co-workers examined various CDC reactions using diverse substrates[8] and reported that CuBr catalyzed alkynylation of N-aryltetrahydroisoquinolines in the presence of tert-butyl hydroperoxide (TBHP) to give the C-1-substituted products.[8] A similar CDC reaction of N,N-dimethylbenzyl amine exclusively gives the N-methyl substituted product. Xiaonian Li and Xiaoliong Xu studied CDC reactions of unactivated aliphatic tertiary amines and terminal alkynes using CuI–diethyl azodicarboxylate (DEAD).[9] These reactions furnish the methyl alkynylated products in excellent yield.[9]

C-1 alkynylation

Materials and Methods

Column chromatography was carried out on silica gel. Reagents were purchased from commercial suppliers and used without further purification unless stated. N-methyltetrahydroisoquinoline was distilled before use. 1 H NMR (300 MHz) spectra were recorded in CDCl₃. Chemical shifts are reported in ppm using TMS as internal standard.

Discussion

In case of N-methyltetrahydroisoquinoline, which has a rigid structure, it is expected that the proton loss to form the iminium ion may occur preferentially from the C-1 position instead of the methyl group. We have carried oxidative coupling between N-methyltetrahydroisoquinoline and phenylacetylene . The reaction was performed by taking the amine–alkyne–DEAD–CuI mixture in the ratio 1:1.5:1.1:0.05 in THF as solvent and the reaction mixture was stirred at room temperature for six hours. Workup and purification by column chromatography gave the C-1 alkynylated amine as the only product in 82% yield. Although three different kinds of hydrogens alpha to the nitrogen atom are present in N-methyltetrahydroisoquinoline, alkynylation occurred mainly at the benzylic (C-1) position. In the absence of copper catalyst no reaction was observed. The reaction conditions were optimized by using different solvents and copper catalysts. CuBr and CuCl also afford the C-1 alkynylated product in good yield after prolonging the reaction time. A small decrease in the yield was observed with solvents like dichloromethane, acetonitrile, toluene and diethyl ether. Thus, the best result was obtained with CuI (5 mol%) using THF as solvent.

Table 1. Optimization of reaction conditions

Entry	solvent	catalyst	mol %	yield(%)
1	THF	-		0

2	DCM	CuI	5	78
3	Acetonitrile	CuI	5	75
4	Toluene	CuI	5	76
5	Et$_2$O	CuI	5	73
6	**THF**	**CuI**	**5**	**82**
7	THF	CuBr	5	80
8	THF	CuBr	10	82
9	THF	CuCl	5	70
10	Hexane	CuI	5	54

Under the optimized conditions, differently substituted acetylenes were successfully reacted with N-methyltetrahydroisoquinoline and the corresponding C-1-alkynylated products were obtained in good yields (Table 2, entries 1–4). An aliphatic alkyne also reacted smoothly (Table 2, entry 5). In none of the reactions was the N-methyl alkynylated product observed.

Table 2. Oxidative coupling of *N*-methyl tetrahydroisoquinolines and alkynes using CuI/ DEAD

entry	**Amine**	**Alkyne**	product	yield(%)
1		Ph———H		82

2				70
3				80
4				82
5				81

Conclusion and Significance

Reactions of N-methyltetrahydroisoquinolines and alkynes using CuI–DEAD mainy afford C- 1-substituted propargylic amines and these were further reduced to phenethylisoquinoline alkaloids. Further investigations of these unique CDC reactions using substrates other than alkynes are in progress.

References

1. Ozturk, T. The Alkaloids, Vol. 53; Cordell, G. A., Ed.; Academic Press: New York, 2000, 120. (b) Wu, W.; Beal, J. L.; Fairchild, E. H.; Doskotch, R. W. J. Org. Chem. 1978, 43, 580.
2. (a) Scott, J. D.; Williams, R. M. Chem. Rev. 2002, 102, 1669. (b) Chrzanowska, M.; Rozwadowska, M. D. Chem. Rev. 2004, 104, 3341. (c) Pedrosa, R.; Andre, C. J. Org. Chem. 2001, 66, 243; and references therein.

3. Byrator, E.; Sasse, B. C.; Stark, H.; Schneider, G. ChemBioChem 2005, 6, 997.

4. (a) Shamma, M. The Isoquinoline Alkaloids, Vol. 25; Blomquist, A. T.; Wasserman, H., Eds.; Academic Press: New York, 1972, Chap. 24, 458. (b) Battersby, A. R.; Bradbury, R. B.; Herbert, R. B.; Munro, M. H. G.; Ramage, R. J. Chem. Soc., Perkin Trans. 1 1974, 1394.

5. (a) Cass, L. J.; Frederik, W. S. Am. J. Med. Sci. 1963, 246, 550. (b) Meyers, A. I.; Dickman, D. A.; Boes, M. Tetrahedron 1987, 43, 5095.

6. For representative papers, see: (a) Miyaura, N.; Suzuki, A. Chem. Rev. 1995, 95, 2457. (b) Kotha, S.; Lahiri, K.; Kashinath, D. Tetrahedron 2002, 58, 9633. (c) Nicolaou, K. C.; Bulger, P. G.; Sarlah, D. Angew. Chem. Int. Ed. 2005, 44, 4442. (d) Dyker, G. Handbook of C-H Transformation; Wiley-VCH: Weinheim, 2005. (e) Zapf, A.; Beller, M. Chem. Commun. 2005, 431. (f) Herreries, C.; Yao, X.; Li, Z.; Li, C.-J. Chem. Rev. 2007, 107, 2546.

7. (a) Li, C.-J. Acc. Chem. Res. 2009, 42, 335. (b) Scheuermann, C.-J. Chem. Asian J. 2010, 5, 436. (c) Yeung, C. S.; Dong, V. M. Chem. Rev. 2011, 111, 1215.

8. (a) Li, Z.-P.; Li, C.-J. J. Am. Chem. Soc. 2004, 126, 11810. (b) Li, Z.-P.; Li, C.-J. Org. Lett. 2004, 6, 4997. (c) Li, Z.-P.; Li, C.-J. J. Am. Chem. Soc. 2005, 127, 3672. (d) Li, Z.-P.; Li, C.-J. J. Am. Chem. Soc. 2005, 127, 6968. (e) Zhang, Y.; Li, C.-J. J. Am. Chem. Soc. 2006, 128, 4242. (f) Murahashi, S.-I.; Nakae, T.; Terai, H.; Komiya, N. J. Am. Chem. Soc. 2008, 130, 11005.

9. (a) Xu, X.; Li, X. Org. Lett. 2009, 11, 1027. (b) Xu, X.; Li, X.; Ma, L.; Ye, N. J. Am. Chem. Soc. 2008, 130, 14048.

ISBN-13: 978-1533168078
ISBN-10: 1533168075

Chemistry of Industrial Globalization, Environmental Pollution and its Chem-Biological Significance

February 2016

Industrial Globalization and Environmental Awareness

Proceedings of the NATIONAL SEMINAR held by
Department of Chemistry, Government PG College, Ambala Cantt - Haryana - India

A BRIEF REVIEW ON GREEN CHEMISTRY

Harish Kumar Soni*, Parvesh Gupta**

*Assistant Professor, Dept. of Chemistry, Govt.P.G. College, Ambala Cantt.
**Assistant Professor, Dept. of Chemistry, R.G. Govt. College, Saha
Email: *Harishkumar484@Yahoo.Com, **Parveshguptaa@Gmail.Com

Abstract

The beginning of green chemistry is frequently considered as a response to the need to reduce the damage of the environment by man-made materials and the processes used to produce them. A quick view of green chemistry issues in the past decade demonstrates many methodologies that protect human health and the environment in an economically beneficial manner. This review emphasize on principle, on human health effects and ecological impacts for a wide variety of individual chemicals and chemical classes, chemists can make informed methodology and recent applications of green chemistry. While it has already been mentioned that nothing is truly environmentally benign, there are substances that is known to be more toxic to humans and more harmful to the environment than others. By using the extensive data available choices as to which chemicals would be more favorable to use in a particular synthesis or process. Simply stated, Green Chemistry is the use of chemistry techniques and methodologies that reduce or eliminate the use or generation of feedstock, products, by-products, solvents, reagents, etc., that are hazardous to human health or the environment. Green Chemistry is an approach to the synthesis, processing and use of chemicals that reduces risks to humans and the environment.

1. Introduction

Over the past few years, the chemistry community has been mobilized to develop new chemistries that are less hazardous to human health and the environment. This new approach has received extensive attention and goes by many names including Green Chemistry, Environmentally Benign Chemistry, Clean Chemistry, Atom Economy and Benign by Design Chemistry. The study of the organic reactions from the point of view of its greenness must have in mind first of all that a general synthetic method must be based on complete and efficient conversions of well defined selectivity and that greenness is more a term for comparison than an absolute kind of qualification. In order to evaluate the greenness of a particular process attention must be paid in the first instance to issues related to safety, health and protection of the environment, due to reactants (substrates and reagents), auxiliaries (mainly solvents) and waste. This enumeration is obviously incomplete, but can be useful at present. The question about how green a reaction is most frequently refers to a particular conversion, to the comparison between two or more alternative processes for the same synthetic target, or between the synthetic pathways for the manufacture of alternative compounds. The study of the greenness of the organic reaction is completed by a short overview of recent contributions indented to achieve efficient, safe and clean conversions that are susceptible to becoming general synthetic procedures[1] With the increasing concerns about the environmental protection, synthesis of organic compounds from raw materials through a Green Chemistry procedure is desirable. Certainly the area of environmentally benign solvents has been one of the leading research areas of Green Chemistry with great advances seen in aqueous (biphase) catalysis [2, 3] and the use of supercritical fluids [4] in chemical reactions. While the greenness of ionic liquids [5, 6] and fluorous media [7] will ultimately depend on their individual properties with respect to health and the environment, the sustainability of new bio based solvents [8] has to be proven as well. There has been a renewed focus on the age-old pursuit of the organic chemist to design and successfully apply the ideal synthesis in terms of efficiency, with atom[9-11]and step economy[11] being a major goal. New catalytic processes continue to emerge to advance the goals of Green Chemistry, while techniques such as microwave[12-14] and ultrasonic synthesis[15] as well as in situ spectroscopic methods[16-17] has been used extensively, leading to spectacular results. These research areas are a glimpse of some of the many topics directly relevant to Green Chemistry being pursued by researchers around the world. The development of asymmetric reactions stereo selective formation of C-C bond based on green protocol is also of paramount interest.

1.1 Definition:

"The invention, design and application of chemical products and processes to reduce or to eliminate the use and generation of hazardous substances" [18] While this short definition appears straightforward, it marks a significant departure from the manner in which environmental issues have been considered or ignored in the up-front design of the molecules and molecular transformations that are at the heart of the chemical enterprise. Looking at the definition of Green Chemistry, the first thing one sees is the concept of invention and design. By requiring that the impacts of chemical products and chemical processes are included as design criteria, the definition of Green Chemistry inextricably links hazard considerations to performance criteria. Another aspect of the definition of Green Chemistry is found in the phrase "use and generation". Rather than focusing only on those undesirable substances that might be inadvertently produced in a process; Green Chemistry also includes all substances that are part of the process. Therefore, Green Chemistry is a tool not only for minimizing the negative impact of those procedures aimed at optimizing efficiency, although clearly both impact minimization and process optimization are legitimate and complementary objectives of the subject. Finally, the definition of Green Chemistry includes the term "hazardous". It is important to note that Green Chemistry is a way of dealing with risk reduction and pollution prevention by addressing the intrinsic hazards of the substances rather than those circumstances and conditions of their use that might increase their risk. The definition of Green Chemistry also illustrates another important point about the use of the term "hazard". This term is not restricted to physical hazards such as explosiveness, flammability, and corrodibility, but certainly also includes acute and chronic toxicity, carcinogenicity, and ecological toxicity. Furthermore, for the purposes of this definition, hazards must include global threats such as global warming, stratospheric ozone depletion, resource depletion and bioaccumulation, and persistent chemicals. To include this broad perspective is both philosophically and pragmatically consistent. It would certainly be unreasonable to address only some subset of hazards while ignoring or not addressing others. But more importantly, intrinsically hazardous properties constitute those issues that can be addressed through the proper design or redesign of chemistry and chemicals. Green Chemistry definitions change based upon focus. Green Chemistry is often described within the context of new technologies But Green Chemistry is not beholden to ionic liquids, [19] microwave chemistry, [20] supercritical fluid, [21] biotransformation, [22] fluorous phase chemistry, [23] or any other new technology. Green Chemistry is outside of techniques used but rather resides within the intent and the result of technical application. Some view Green Chemistry as something process chemists do already Good process chemistry. While often enabling "greener" synthesis, good process chemistry is not equivalent to Green Chemistry. A robust, efficiency, and cost5 effective chemical process is likely accepted as good process chemistry. Green Chemistry is not simply good process chemistry. In short, Green Chemistry is neither a new type of chemistry nor an environmental movement, a condemnation of industry, new technology, or "what we do already". Green Chemistry is simply a new environmental priority when accomplishing the science already being performed... regardless of the scientific discipline or the techniques applied. Green Chemistry is a concept driven by efficiency coupled to environmental responsibility. Green Chemistry philosophy [24] provides a design for chemical evolution and a guide for scientists to accomplish sustainable practices during chemical research, development, and manufacturing. It has been proposed that evolution toward Green Chemistry has recently crested a summit [25] and gained momentum enough that general technical exemplification is both imminent and inevitable. A literature search may provide no current alternative with similar efficiency and reduced toxicity, but many do not realize that the simple act of inquiry toward reduced toxicity already indicates a new priority and intent, a higher level of awareness and environmental stewardship, and is Green Chemistry! In some cases a safer reagent will exist. Green Chemistry is defined as environmentally benign chemical synthesis. Green Chemistry may also be defined as the invention, design, and application of chemical products and processes to reduce or eliminate the use and generation of hazardous substances. The synthetic schemes are designed in such a way that there is least pollution to the environment. As on today, maximum pollution to the environment is caused by numerous chemical industries. The cost involved in disposal of the waste products is also enormous. Therefore, attempts have been made to design synthesis for manufacturing processes in such a way that the waste products are minima, they have no effect on the environment and their disposal is convenient. For carrying out reactions it is necessary that the starting materials, solvents and catalysts should be carefully chosen. For example, use of benzene as a solvent must be avoided at any cost since it is carcinogenic in nature. If possible, it is best to carry out reactions in the aqueous phase. With this view in mind, synthetic methods should be designed in such a way that the starting materials are consumed to the maximum extent in the final product. The reaction should not generate any toxic by-products. Since its birth over a decade ago, the field of Green Chemistry has seen rapid expansion, with numerous innovative scientific breakthroughs associated with the production and utilization of chemical products. [26] The concept and ideal of Green Chemistry now goes beyond chemistry and touches subjects ranging from energy to societal sustainability. The key notion of Green Chemistry is ''efficiency'', including material efficiency, energy efficiency, man-power efficiency, and property efficiency (e.g., desired function vs toxicity). Any ''wastes'' aside from these efficiencies are to be addressed through innovative Green Chemistry means. ''Atom-economy'' [27] and minimization of auxiliary chemicals, such as protecting groups and solvents, form the pillar of material efficiency in chemical productions. Green Chemistry is an approach to the design, manufacture and use of chemical products to intentionally reduce or eliminate chemical hazards [28] Goal of Green Chemistry is to create better, safer chemicals while choosing the safest, most efficient ways to synthesize them and to reduce wastes. Chemicals are typically created with the expectation that any chemical hazards can somehow be controlled or managed by establishing "safe" concentrations and exposure limits [28].

2. Bibliographical Sources: Examples of new "greener" synthetic methods with principles of Green Chemistry

The scientific literature is vast and contains a great number of research publications on the principles of Green Chemistry and how these can be applied to organic synthetic routes for "old" and conventional methods. The subjects of Green Chemistry are covered by various internet sites and networks which publish continuously new research.
URL:http://www.organic-chemistry.org/topics/greenchemistry.shtm

1. Transformation of aromatic and aliphatic alcohols in the equivalent carboxylic acids and ketones. Green synthetic method:
Various aromatic, aliphatic and conjugated alcohols were transformed into the corresponding carboxylic acids and ketones in good yields with aq 70% t-BuOOH in the presence of catalytic amounts of bismuth (III) oxide. This method possesses does not involve cumbersome work-up, exhibits chemo selectivity and proceeds under ambient conditions. is The overall method green.[29]

2. Enantioselective Michaell addition :
A highly enantioselective Michael addition of malonates to α,β-unsaturated ketones in water is catalyzed by a primary-s-econdary diamine catalyst containing a long alkyl chain. This asymmetric Michael addition process allows the conversion of various α,β-unsaturated ketones. [30]

3. Organocatalytic direct α-hydroxy phosphonate of aldehydes and ketones:
An organocatalytic, direct synthesis of α-hydroxy phosphonates via reaction of aldehydes and ketones with trimethylphosphite in the presence of catalytic amounts of pyridine 2, 6-dicarboxylic acid in water is simple, cost-effective and environmentally benign. [31]

4. *Intermolecular addition of perfluoroalkyl radicals:*
Intermolecular addition of perfluoroalkyl radicals on electron rich alkenes and alkenes with electron withdrawing groups in water, mediated by silyl radicals gives perfluoroalkyl-substituted compounds in good yields. The radical triggering events employed consist of thermal decomposition of 1,1′azobis(cyclohexanecarbonitrile) (ACCN) or dioxygen initiation.[32]

 5. Practical catalytic method for N-formylation
A simple, practical, and catalytic method for the N-formylation in the presence of molecular iodine as a catalyst under solvent-free conditions is applicable to a wide variety of amines. α-Amino acid esters can be converted without epimerization. [33]

6. Direct oxidation of methyl group in aromatic nucleus

A methyl group at an aromatic nucleus is oxidized directly to the corresponding carboxylic acid in the presence of molecular oxygen and catalytic hydrobromic acid under photoirradiation.[34]

7. Oxidation of sulfides with H_2O_2

Oxidation of sulfides with 30% hydrogen peroxide catalyzed by tantalum carbide provides the corresponding sulfoxides in high yields, whereas niobium carbide as catalyst efficiently affords the corresponding sulfones. Both catalysts can easily be recovered and reused without losing their activity. [35]

8. Borono-Mannich reactions in solvent-free conditions

Borono-Mannich reactions can be performed in solvent-free conditions under microwave irradition with short reaction time. Full conversion of the starting materials towards the expected product was achieved, starting from stoichiometric quantities of reactants, avoiding column chromatography. No purification step other than an aqueous washing was required. [36]

9. Oxidation of alkynes in aqueous media

Oxidation of alkynes using ammonium per sulfate and diphenyl diselenide as catalyst in aqueous media leads to 1, 2-unprotected dicarbonyl derivatives or to hemiacetals starting from terminal alkynes. [37]

10. Mild method for N-formylation in the presence of Indium metal

A simple, mild method for N-formylation in the presence of indium metal as a catalyst under solvent-free conditions are applicable to the chemo selective reaction of amines and α-amino acid esters without epimerization. [38]

2.1 New "Green" Methods in Synthetic Organic Sonochemistry:

Ultrasound-assisted organic synthesis is another "green" methodology which is applied in many organic synthetic routes with great advantages for high efficiency, low waste, low energy requirements. Sonochemistry (in the region of 20 kHz to 1 MHz) has many applications due to its high energy and the ability to disperse reagent in small particles and accelerate reactions [39-40]. Irradiation with high intensity sound or ultrasound, acoustic cavitations usually occurs (growth, and implosive collapse of bubbles irradiated with sound). Experimental results have shown that these bubbles have temperatures around 5000 K, pressures of roughly 1000 atm. These cavitations can create extreme physical and chemical conditions in otherwise cold liquids. [41-43] Also, Sonochemical engineering is a new field involving the application of sonic and ultrasonic waves to chemical processing. Sonochemistry enhances or promotes chemical reactions and mass transfer. It offers the potential for shorter reaction cycles, cheaper reagents, and less extreme physical conditions. Existing literature on sonochemical reacting systems is chemistry-intensive, and applications of this novel means of reaction in environmental remediation and pollution prevention seem almost unlimited and are rapidly growing area. [44]

2.2 Use of precursor materials:

A great attempt has been made to shift the usage of petroleum based products which currently forms 95% of cases as a starting material for various chemicals required in tons per year. On this basis shifting to biomass for such a vast need is the call of an hour. Researchers are now finding new ways of converting biomass into starting material, which in some or the other has been a great success. E.g. converting D-glucose into lactic acid using certain enzymes helps us to prepare aliphatic compounds from lactic acid. On a similar basis E-coli converts D- glucose to catechol which acts as a starting material for aromatic compounds. On the other hand isomaltulose which is widely available in biomass can be converted into glucosylmethyl furfural which can be used for production of many heterocyclic compounds. Besides biomass cash crops is a new hope as ethanol from sugarcane has been derived successfully and now scientists are trying to use this "bio alcohol" as a source of vehicle for future. Exhaust from Corn plant has been successfully utilized for preparing bio-degradable plastic [45].

3. Effect of carbon footprint:

It is a point of concern that about 1/7th of the total energy production (including electricity, petroleum and its related products, coal, and wood) are utilized in the production of chemicals by industries. Which is not only a large share of energy consumption but plays a big role in environment related hazards including global warming. Here green chemistry can play a major role in reducing carbon footprint. Minimizing the energy requirements of industries by maximizing the efficiency of chemical conversion and decreasing the activation energy of the reactions by using recyclable catalysts can cut off the energy requirement of industries by half or even more. Eliminating the use of energy consuming steps like distillation, crystallization, sublimation, ultra filtration etc. and incorporation of microwave energy which aims to achieve a high temperature at much faster rates and also utilization of ultrasonic energy for certain reaction can eventually solve this problem [45]. Besides the above mentioned "pillars of green chemistry" some other points that can also been incorporated as the supports of green chemistry are as follows: a) Use of "light" as a carrier of electrons which can eventually reduce the usage of other chemical agents which act as a carrier of electron and is obtained as waste products at the end of a redox reaction. b) Eliminating the un-necessary use of protection- deprotection methodologies. c) Replacement of soluble Lewis acids by mesoporous solids containing bound sulphonates in green synthesis. d) Utilization of milder reaction conditions for carrying out a chemical reaction.

Environmental concerns in synthetic chemistry have led to a reconsideration of reaction methodologies. This has resulted in investigations into atom economy, [46] the use of supercritical CO_2, [47] ionic liquids, [48] and other procedures to reduce the disposal problems associated with most chemical reactions. One obvious route to reduce waste entails generation of chemicals from reagents in the absence of solvent. Therefore the design of green processes with no use of hazardous and expensive solvents, e.g., "solvent-free" reactions, has gained special attention from synthetic organic chemists. [49] As a result, many reactions are newly found to proceed cleanly and efficiently in the solid state or under solvent-free conditions. [50] Less chemical pollution, lower expenses, and easier procedures are the main reasons for the recent increase in the popularity of solvent-free reactions. While an obvious approach to chemical synthesis, there are many problems associated with this approach, the chief of which is the role of diffusion/interactions between reactants. Further, it is never clear that the reactions in the solid state will generate the same products as those found in the presence of solvents. [51] Generally, Michael additions are conducted in a suitable solvent in the presence of a strong base either at room temperature or at elevated temperatures. [52] Due to the presence of the strong base, side reactions such as multiple condensations, polymerizations, rearrangements and retro-Michael additions are common. These undesirable side reactions decreases the yields of the target adduct and render their purification difficult. Better results can be obtained by employing weaker bases such as piperidine, quaternary ammonium hydroxide, tertiary amines etc. [53] there have been some reports on Michael reactions catalyzed by potassium carbonate in organic solvents, [54] and water in the presence of surfactants [55] or phase-transfer catalysts. [56] To a large extent, mild bases restrain the formation of side products, thus improving the yield of the desired Michael adducts. Recently, non-conventional procedures like conducting the reaction on the surface of a dry medium or under microwave irradiation [57] were found to facilitate the Michael reaction. For the purposes of eco-friendly "Green Chemistry", a reaction should ideally, be conducted under solvent-free conditions with minimal or no side-product formation and with utmost atom-economy.

4. Green Chemistry Metrics:

The Environmental factor E for Waste in Chemical Reactions Green Chemistry introduced various general metrics to give quantitative meaning of chemical processes. The environmental E-factor was established as the indicator of mass waste per unit of product in the chemical industry. The E-factor can be made as complex and thorough or as simple as required. Assumptions on solvent and other factors can be made or a total analysis can be performed.

The E-factor calculation is defined by the ratio of the mass of waste (kg) per unit of product in kilograms:

E-factor = total waste (kg) / product (kg)

The Green Chemistry metric is very simple to understand and to use. It highlights the waste produced in the process as opposed to the reaction, thus helping those who try to fulfill one of the twelve principles of green chemistry to avoid waste production. The environmental E-factors ignore recyclable factors such as recycled solvents and re-used catalysts, which obviously increases the accuracy but ignores the energy involved in the recovery. Roger A. Sheldon took his publications one stage further and produced the following Table for E-Factors across the chemical industry [57a-57e]

Industry sector	Annual production (t)	E-factor	Waste produced
Oil refining	10^6-10^8	Ca. 0.1	10^5-10^7
Bulk chemicals	10^4-10^6	<1–5	10^4-5×10^6
Fine chemicals	10^2–10^4	5–50	5×10^2–5×10^5
Pharmaceuticals	10–10^3	25–100	2.5×10^2–10^5

5. Microwave Applications for Green Chemistry Synthesis

Microwave applications in organic synthesis are not something new. But it is interesting to realize the potential of this synthetic method with low energy requirements, less waste, no use of solvent. The principles of Green chemistry apply to most of the synthetic routes with microwave irradiation. Microwave-assisted eco-friendly organic synthesis has become a new trend with many applications in synthesizing organic chemicals. Organic reactions under the microwave irradiation have many advantages compared to the conventional reactions which need very high temperatures. Microwave assisted reactions are "cleaner", last only very few minutes, have high yield and produce minimum waste.[58] Microwave assisted organic synthesis has become an expanding field in synthetic research. New publications cover the many aspects of this "greener" technique and its practical applications [59-64] the scientific literature is full of new research papers on microwave reaction mechanisms and applications. [65-69]

CONCLUSION:

Goal of Green Chemistry is to create better, safer chemicals while choosing the safest, most efficient ways to synthesize them and to reduce wastes. Chemicals are typically created with the expectation that any chemical hazards can somehow be controlled or managed by establishing "safe" concentrations and exposure limits. A good flow of knowledge between the Industries and research institutions/ universities undergoing such types of research topics will not only enable us to expand our knowledge but it would also help to protect the environment. Government should also make some strict rules in governing the industries to use eco-friendly ways of production. It may be incorporated in the syllabus of UG and PG programme about green chemistry to educate young students.

References:

1. Clark, J.H.Green Chem.1999, 1, 1.
2. Li, C. Chem. Re .1993, 93, 2023 and 2005, 105, 3095.
3. Aqueous Organometallic Chemistry and Catalysis; Horvath, I. T., Joo, F., Eds.; Kluwer   Academic Publishers: Dordrecht, 1995.
4. Jessop, P.; Leitner, W. Chemical Synthesis Using Supercritical Fluids; WileyVCH: Weinheim,   1999.
5. Welton, T. Chem. Re .1999, 99, 2071.
6. Earle, M.; Seddon, K. Pure Appl. Chem. 2000, 72, 1391.
7. Horvath, I. T. Acc. Chem. Res. 1998, 31, 641.
8. Dale, B. J. Chem. Technol. Biotechnol.2003, 78, 1093.
9. Trost, B. Science 1991, 254, 1471.
10. Trost, B. Acc. Chem. Res. 2002, 35, 695.
11. Wender, P. A.; Croatt, M. P.; Witulski, B. Tetrahedron 2006, 62, 7505.
12. Bose, A. K.; Manhas, M. S.; Ganguly, S. N.; Sharma, A. H.; Banik,B. K. Synthesis 2002, 1578.
13. Nuchter, M.; Ondruschka, B.; Bonrath, W.; Gum, A. Green Chem.2004, 6, 128.
14. Topics in Current Chemistry: Microwa e Methods in Organic Synthesis; Larhed, M., Olofsson, K., Eds.; Springer: Berlin, 2006.
15. Mason, T. J. Sonochemistry; Oxford University Press: Oxford, 1999.
16. Chalmers, J. M. Spectroscopy in Process Analysis; CRC Press: Boca Raton, FL, 2000.
17. In situ NMR Methods in Catalysis; Bargon, J., Kuhn, L. T., Eds.; Springer: Dordrecht, 2007.
18. P.T. Anastas and J. C.Warner.Green Chemistry: Theory and Practice.Oxford, Science Publications, Oxford (1998).
19. Seddon, K. Green Chem. 2002, 4, G35. Rogers, R.D.; Seddon, K .Ionic Liquids: Industrial Applications for Green Chemistry; ACS Ser.818; Oxford University Press: Oxford, UK,2002.Wasserscheid, P.; Welton,T. Ionic liquids in Synthesis;Wiley-VCH: Weinheim, 2003 .Welton, T.Coord .Chem. Rev. 2004, 248, 2459, Chauhan , S.M.S.; Chauhan, S.; Kumar , A.; Jain, N. Tetrahedron 2005, 61, 1015.

20. Caddick, S. Terahedron 1995, 51, 10403. Strauss, C.R.; Trainor, R. W. Aust. J.Chem. 1995, 48, 1665. Loupy, A.; Petit, A.; Hamelin, J.; Texier-Boullet, F.; Jacquault, P.; Mathe, D.Synthesis 1998, 1213.Wathey, B.; Tierney, J.; Lidstorm, P.; Westman, J.Drug Discovery Today 2002, 7, 373.Loupy,A.Microwaves in Organic Synthesis,Wiley-VCH: Weinheim, 2002. Nuchter, M.; Ondruschka, B.; Bonrath, W.; Gum, A. Green Chem.2004, 6,128. Hayes ,B.Aldrichimica Acta 2004,37,66.Stadler,A.; Kappe, O.C. Microwave-Assisted Solid-Phase Synthesis;Blackwell Ltd.; Oxford, 2005. Tierney, J.; Lidstrom, P. Microwave Assisted Organic Synthesis; Blackwell Ltd.; Oxford, 2005.

21. Jessop, P.G.; Leitner, W.Chemical Synthesis Using Supercrictical Fluids;Wiley VCH:.Weinheim, Germany,1999. Leitner, W.Acc.Chem .Res. 2002, 35, 746. Poliakoff, M.; Ross, S.K.; Sokolova, M.; Ke, J.; Licence, P.Green Chemistry 2003, 5, 99. Beckman, E.J.J.Supercrit.Fluids 2004, 28,121.

22. Tao,J.;Yazbeck,D.;Martinez,C.A.;Hu,S.Tetrahedron:Assymetry 2004,15,2757.

23. Curran, D.P.Angew.Chem.; Int.Ed.1998, 37, 1174.Curran, D.P.; Lee, Z.Green Chem.2001, G3. Gladysz, J.A.; Curran, D.P.; Horvath, I.T .Handbook of Fluorous Chemistry; Wiley-VHS: 2004.

24. Anastas, P.T.; Warner, J.C.Green Chemistry Theory and Practice; Oxford University Press: New York, 1998. (b) Woodhouse, E.J.; Chemical States; Casper, M., Ed.; Routledge: New York, 2003. (c)Tucker, J.L.Org.Process Res.Dev.2006, 10, 315.

25. Cue, B.W.11th Annual Green Chemistry and Engineering Conference, Washington DC, June, 2007.

26. P. T. Anastas and J. C. Warner, Green Chemistry Theory and Practice, Oxford University Press, New York, 1998.

27. B. M. Trost, Science, 1991, 254, 1471.

28. Anastas and Warner, Green Chemistry: Theory and Practice, 1998

29. Malik P., D. Chakraborty D. Bismuth (III) oxide catalyzed oxidation of alcohols with tert-butyl hydroperoxide. , Synthesis, 2010, 37363740.

30. Z. Mao, Y. Jia, W. Li, R. Wang, J. Org. Chem., 2010, 75, 7428-7430.

31. F. Jahani, B. Zamenian, S. Khaksar, M. Taibakhsh, Synthesis, 2010, 3315-3318.

32. S. BarataVallejo, A. Postigo, J. Org. Chem., 2010, 75, 6141-6148.

33. J.-G. Kim, D. O. Jang, Synlett, 2010, 2093-2096.

34. S.-I. Hirashima, A. Itoh, Synthesis, 2006, 1757-1759.

35. M. Kirihara, A. Itou, T. Noguchi, J. Yamamoto, Synlett, 2010, 1557-1561.

36. P. Nun, J. Martinez, F. Lamaty, Synthesis, 2010, 2063-2068.

37. S. Santoro, B. Battistelli, B. Gjoka, C.-w. S. Si, L. Testaferri, M. Tiecco, C. Santi, Synlett, 2010, 14021406.

38. J.-G. Kima, D. O. Jang, Synlett, 2010, 1231-1234

39. Martyn Poliakoff, J. Michael Fitzpatrick, Trevor. Farren Paul I, Anastas, Science, 2002, 2, 297.

40. See Resources section plus visit http://www.epa.gov/ greenchemistry/index.html

41. Cintas P, Luche J-L. Green Chemistry: The sonochemical approach. Green Chem 1:115-125, 1999. 32. Mason TJ. Ultrasound in synthetic organic chemistry. Review. Chem Soc. Rev. 26:443-451, 1997. 33. Cravotto G, Cintas P. Power ultrasound in

organic synthesis: moving cavitational chemistry from academia to innovative and large-scale applications. Chem. Soc. Rev. 35:180-196, 2006.

42. Tanaka, K.; Toda, F. Chem. Rev. 2000, 100, 1025. (b) Bradley, D. Chem. Br. 2002, 42 (Sept).

43. Seddon, K. R. J. Chem. Technol. Biotechnol.1997, 68, 351.

44. Mason TJ. Sonochemistry. Oxford University Press, Oxford, 1999; Adewuyi YG. Sonocemistry: Environmental science and engineering applications. Industr Eng Chem Res 40(22): 4681-4715, 2001.

45. Robert thornton Morrison, Robert Neilson boyd and Saibal Kanti Bhattacharjee. "organic chemistry" seventh edition pg. no. 1419- 1427.

46. Trost, B. M. Acc. Chem. Res. 2002, 35, 386.

47. Wells, S.; DiSimone, J. M. Angew.Chem., Int. Ed. 2001, 40, 519.

48. Tanaka, K. Solvent-Free Organic Synthesis; Wiley-VCH: Weinheim, Germany, 2003. (b) Tanaka, K.; Toda, F. Chem. Re. 2000, 100, 1025- 1074.

49. Zhao, J.L.; Liu, L.; Sui, Y.; Liu, Y. L.; Wang, D.; Chen, Y. J. Org. Lett. 2006, 8, 6127- 6130. Castrica, L.; Fringuelli, F.; Gregoli, L.; Pizzo, F.; Vaccaro,L. J. Org. Chem. 2006, 71, 9536- 9539. Azizi, N.; Aryanasab, F.; Saidi, M. R. Org. Lett. 2006, 8, 5275- 5277. Lofberg, C.; Grigg, R.; Whittaker,M. A.; Keep, A.; Derrick, A. J. Org. Chem. 2006, 71, 8023- 8027. Hosseini-Sarvari, M.; Sharghi, H. J. Org. Chem. 2006, 71, 6652- 6654. Mojtahedi, M. M.; Abaee, M. S.; Heravi, M. M.; Behbahani, F. K. Monatsh.Chem. 2007, 138, 95- 99. Choudhary, V. R.; Jha, R.; Jana, P. Green Chem. 2007, 9, 267 272.

50. Jung M E 1993 Comprehensive organic synthesis (eds) B M Trost and I Fleming (Oxford: Pergamon Press) vol. 4, p. 1–68.

51. Bergmann E D, Ginsburg D and Pappo R Organic React. 1959, 10, 179.

52. House H O Modern synthetic reactions (ed.) W A Benjamin (New York: Amsterdam) 1972, p. 595.

53. Rosnati V, Saba A and Salimbeni A Tetrahedron Lett. 1981, 22, 167.

54. Toda F, Takumi H, Nagami M and Tanaka K Hetrocycles, 1998, 47, 469

55. Bram G, Sansoulet J, Galons H and Miocque M Synth.Commun. 1988, 18, 367; (b) Kim D Y, Huh S C and Kim S M Tetrahedron Lett. 2001, 42, 6299; (c) Dere R T, Pal R R, Patil P S and Salunkhe M M Tetrahedron Lett. 2003, 44, 5351.

56. Ranu B C and Bhar S Tetrahedron, 1992, 48,1342; (b) Ranu B C, Bhar S and Sarkar D C Tetrahedron Lett. 1991, 32 , 2811.

57. 57a. Romanova N N, Gravis A G, Sharidullina G M, Leshcheva I F and Bundel Y G Mendeleev Commun. 1997, 213.
57b.Lapkin A, Constable D. Green Chemistry Metrics. Measuring and Monitoring Sustainable Processes, Wiley, West Sussex, UK, 2008.
57c. Sheldon RA. Atom efficiency and catalysis in organic synthesis. Pure Appl Chem 72(7):1233-1246, 2000.
57d.Sheldon Roger A.: ChemInform Abstract: Atom Efficiency and Catalysis in Organic Synthesis. ChemInform 32, 2001..
57e.Sheldon Roger A.: Utilisation of biomass for sustainable fuels and chemicals: Molecules, methods and metrics. Catal Today 167, 3, 2011.

58. Lidstrom P, Tierney J, Wathey B, and Microwave assisted organic synthesis: a review. Tetrahedron 57:9225-9283, 2001.
59. Desai KR, Kanetkar VR. Green Chemistry Microwave Synthesis. Global Media, New York, 2010.
60. One Hundred Reaction Procedures. (Tetrahedron Organic Chemistry). Elsevier, New York, 2006.
61. Loupy A (Ed). Microwaves in Organic Synthesis. Wiley-VCH, West Sussex, UK, 2002 (1st), 2006 (2nd ed).
62. Kappe CO, Dallinger D, Murphree SS. Practical Microwave Synthesis for Organic Chemistry. Strategies, instruments, and protocols. Wiley-VCH, West Sussex, UK, 2009.
63. Tierney JP, Lidstrom P. Microwave Assisted Organic Synthesis. WileyBlackwell, London, 2005.
64. Larherd M, Olofsson (Eds). Microwave Methods in Organic Synthesis. Springer, Berlin, 2006.
65. Roberts BA, Strauss CR. Toward rapid "green" predictable microwaveassisted synthesis. Review. Acc. Chem. Res. 38:653-661, 2005.
66. Nuchter M, Ondruschka B, Bonrath W, Gum A. Microwave assisted synthesis- a critical technology overview. Green Chem. 6:128-141, 2004.
67. Santagada V, Frecentese F, Perissuti E, et al. Microwave assisted synthesis: a new technology in drug discovery. Review. Mini Rev Med Chem 9:340-358, 2009.
68. Martinez-Palou R. Microwave-assisted synthesis using ionic liquids. Review. Mol Divers 14:3-25, 2010.
69. Alcazar J, Oehlrich D. Recent applications of microwave irradiation to medicinal chemistry. Review. Future Med Chem 2:169-178, 2010.

ISBN-13: 978-1533168078
ISBN-10: 1533168075

Chemistry of Industrial Globalization, Environmental Pollution and its Chem-Biological Significance

February 2016

Industrial Globalization and Environmental Awareness

Proceedings of the NATIONAL SEMINAR held by

Department of Chemistry, Government PG College, Ambala Cantt - Haryana - India

Indoor Pollutants-A threat to Human Health

Rajeev Sharma[1], Jyoti Shah[2] and Sanjeev Kumar[2]

[1]Post Graduate Department of Chemistry, Multani Mal Modi College, Patiala-147001
[2]Environment Research Laboratory, Multani Mal Modi College, Patiala-147001
E mail- rajeev.sharma00@yahoo.com

Abstract

Indoors, the places like home, offices, clubs, restaurants etc. are generally thought to be free from pollutants and safer places for humans. But it has been realized that tobacco smoke, combustion processes like stove and oven operations, pesticides spraying pesticide cool or film burning and emanations from certain types of particle boards, cements, lime, wood and other building materials are often the most significant determinants of indoor air quality. Offices presents their own set of pollutants and difficulties. The chemicals like carbon, formaldehyde, some organic compounds, nitrogen oxides, oxides of carbon, inhalable particle, like asbestos, tobacco smoke, ozone etc. are some chemical pollutants and various kind of pests, mites, and disease causing organisms like viruses, bacteria, fungal spores, amoebae, actinomycetes, endotoxins etc are some biological pollutants found to be present in an indoors environment. These indoors pollutants are posing a threat to human health. More susceptible are the infants, children, old people, and pregnant women who remain inside the home. Adverse effects are seen during rainy season and winters.

In the present paper we aimed to discuss the various chemical and biological pollutants in the indoor places, their effects on health, environment and their control.

Introduction

Due to civilization most part of world's population live in modern single and multifamily dwellings. People work in modern office buildings, institutions and carry various activities in built environment. Indoors pollution may be defined as any physical, chemical and biological characteristics of air in indoor environment e.g. within a home, office, building, an institution or commercial facility. To fulfill the cooking and heating needs people use crude biomass fuels (crop residues, animal dung and wood) inside their shelters, which generates smoke that adversely affect the human health. Due to increased industrialization there is a significant increase in use of coal in indoors during winter season (Zhang and Smith, 2003). In rural and urban areas, different conditions are responsible for indoor air pollution. In developed countries indoor air pollution is a main topic of concern, where sometimes energy efficiency improvements make houses relatively airtight, reducing ventilation and raising pollutant levels. Rural areas in developing countries face greatest threat from indoor pollution. In developing countries some 3.5 billion people continue to depend on traditional fuels such as charcoal, firewood, and cowdung used for cooking and heating. This results in increase in concentration of indoor pollution (smoke and other air pollutant) in households at alarming rate. Sometimes carbon monoxide (CO) poisoning & cases have been reported due to improper use and or inadequate ventilation of appliances. Fossil fuels such as natural gas, liquified petroleum gas, heating oil (petroleum product) and electricity is used for cooking and space heating (Zhang and Smith, 2003).

In 1992, in developing countries the World Bank designated indoor air pollution as one of the most critical global environmental problems. Women and children are the most vulnerable groups that are exposed to the smoke because they spend more time indoors.

Sources of Indoor Air Pollutants: It is very important to know the source of Indoor pollutants, how much of a given pollutant it emits, how hazardous those emissions are, occupant proximity to the emission source, and the ability of the ventilation system to remove the contaminant. In some cases, factors such as the age and maintenance history of the source are significant. Sources of indoor air pollution may include (OSHA, 2011):

Site or Location of building: Location of building may be the cause of indoor air pollution e.g. highways or busy thoroughfares may be sources of particulates and other pollutants in nearby buildings. Buildings sited on land where there was prior industrial use, waste dumping site or where there is a high water table may result in leaching of water or chemical pollutants into the building (OSHA, 2011).

Design of building: Design and construction pattern influence and contribute to indoor air pollution. Poor foundations, roofs, facades, and window and door openings may allow pollutant or water intrusion. Outside air intakes placed near sources where pollutants are drawn back into the building (e.g., idling vehicles, products of combustion, waste containers, etc.) or where building exhaust reenters into the building can be a constant source of pollutants. Buildings with multiple tenants may need an evaluation to ensure emissions from one tenant do not adversely affect another tenant. (OSHA, 2011)

Renovation Activities: When painting and other renovations are being conducted, dust or other particles may be the source of pollution. By products of the construction materials are sources of pollutants that may circulate through a building. Isolation by barriers and increased ventilation to dilute and remove the contaminants are recommended (OSHA, 2011).

Local exhaust and ventilation: Kitchens, laboratories, maintenance shops, parking garages, beauty and nail salons, toilet rooms, trash rooms, soiled laundry rooms, locker rooms, copy rooms and other specialized areas may be a source of pollutants when they lack adequate local exhaust ventilation (OSHA, 2011).

Materials of building: Disturbing thermal insulation or sprayed-on acoustical material, or the presence of wet or damp structural surfaces (e.g., walls, ceilings) or non-structural surfaces (e.g., carpets, shades), may contribute to indoor air pollution (OSHA, 2011).

Furnishings of building: Cabinetry or furniture made of certain pressed-wood products may release pollutants into the indoor air (OSHA, 2011).

Maintenance of building: Workers in areas in which pesticides, cleaning products, or personal-care products are being applied may be exposed to pollutants. Allowing cleaned carpets to dry without active ventilation may promote microbial growth (OSHA, 2011).

Occupant Activities: Building occupants may be the source of indoor air pollutants. Such pollutants include perfumes or colognes (OSHA, 2011).

Common Pollutant Categories

Although there are numerous indoor air pollutants that can be spread through a building, they typically fall into three basic categories: biological, chemical, and particle.

Biological

Excessive concentrations of bacteria, viruses, fungi, dust mites, animal dander, and pollen may result from inadequate maintenance and housekeeping, water spills, inadequate humidity control, condensation, or water intrusion through leaks in the building envelope or flooding.

Chemical

Sources of chemical pollutants (gases and vapors) include emissions from products used in the building (e.g., office equipment; furniture, wall and floor coverings; pesticides; and cleaning and consumer products), accidental spills of chemicals, products used during construction activities such as adhesives and paints, and gases such as carbon monoxide, formaldehyde, and nitrogen dioxide, which are products of combustion.

Particle (Non-biological)

Particles are solid or liquid, non-biological, substances that are light enough to be suspended in the air. Dust, dirt, or other substances may be drawn into the building from outside. Particles can also be produced by activities that occur in buildings such as construction, sanding wood or drywall, printing, copying, and operating equipment.

Major indoor pollutants

There are various pollutants that are generated in doors. It has been reported that gas stove emits nitrogen dioxide (NO_2) (Basu and Samet, 1999). Particulate matter (PM), CO, eye irritation volatile organic compounds (e.g. aldehydes) and carcinogenic compounds such as benzene and 1, 3-butidiene and polycyclic aromatic hydrocarbons generated by gaseous fuels (Zhang et al., 2000; Zhang and Smith, 1999) Synthetic materials and chemical products (Table 1) have been extensively used in modern airtight buildings, which resulted in elevated concentrations of volatile organic compounds (VOCs) (e.g. phthalate plasticizers and pesticides) and human bioeffluents. In the last three decade this is the major contributing factor to illness called as 'sick building syndrome' (Apte et al., 2000).

Table 1. Major pollutants produced from indoors pollutants (Zhang and Smith, 2003).

Pollutants	Major indoors sources
Fine particles	Fuel/tobacco combustion, cleaning, cooking
Carbon monoxide (CO)	Fuel/tobacco combustion
Polycyclic aromatic hydrocarbons	Fuel/tobacco combustion, cooking
Nitrogen oxides	Fuel combustion
Sulphur oxides	Coal combustion
Arsenic and fluorine	Coal combustion
Volatile and semi-volatile organic compounds	Fuel/tobacco combustion, consumer products, furnishings, construction materials, cooking
Aldehydes	furnishings, construction materials, cooking
Pesticides	Consumer products, dust from outside

Asbestos	Remodelling/demolition of construction materials
Lead[a]	Remodelling/demolition of painted surfaces
Biological pollutants	Moist areas, ventilation systems, furnishings
Radon	Soil under building, construction materials
Free radicals and other short lived, highly reactive compounds	Indoor chemistry

[a] pb-containing dust from deteriorating paint is an important indoor pollutant for occupants in many households, but the most critical exposure pathway are not usually through air.

Pollutants generated from burning solid fuels

Incomplete combustion of solid fuels generates products which comprises of CO and fine particles (respirable), volatile organic compounds (VOCs) and semi-volatile organic compounds (SVOCs) (Zhang et al., 2000; Zhang and Smith, 1996; Zhang and Smith 1999; Tsia et al., 2003) Recent report estimated by WHO that indoor smoke from solid fuels ranked one of the top ten risk factors for the global burden of disease which accounts for estimated 1.6 million premature deaths each year (WHO, 2002). It causes risk factor for acute respiratory infections (ARI), chronic obstructive pulmonary disease (COPD) and lung cancer (Coal smoke), cataracts, tuberculosis, asthmatic attacks and adverse pregnancy outcomes (Smith, 2002). Exposure to smoke from solid fuels produces cardiovascular disease.

Environmental Tobacco smoke (ETS)

For more than half a century, tobacco smoke has been accepted and is a well documented cause of ill health. ETS is contains the gases and particles including a wide range of irritating compounds and carcinogens. With limited evidence it has been suggested that breast cancer (Khuder and Simon, 2000) and pulmonary tuberculosis (Alcaide et al., 1996; Altet et al., 1996) have been associated with ETS exposure. In USA estimates indicated that 3000 lung cancer deaths each year can be attributed to passive smoking along with hundreds of thousands of childhood respiratory disease cases (USEPA, 1992). The CDC also estimates that in USA active tobacco smoking causes more than 400,000 deaths each year and results in more than $ 50 billion in direct medical costs annually. Diseases associated with ETS shown table 2.

Table 2: Diseases associated with ETS exposure (Zhang and Smith, 2003)

Disease	Population	Exposure assessment	Number of studies in meta analysis
LRI	Children <3 years of age	Smoking by either parent	24 community and hospital based studies
Asthma	Children >1 year of age	Smoking by either parent	14 case control studies
Otitis media (recurrent)	Children <7 years of age	Smoking by either parent	9 case control, survey, and cohort studies
Ischaemic heart disease	Adults	Lifelong non smokers married to smokers	17 studies
Lung cancers	Adults	Lifelong non smokers with	37 studies

		spouses who currently smoke	
Nasal sinus cancer	Adults	Household exposure to passive smoking	3 studies, no meta study
LBW or SGA	Infants	Prenatal maternal smoking	
Sudden infant death syndrome	Infants <1 year	Prenatal maternal smoking	
LBW or SGA	Infants	ETS exposure of non smoking mothers	
Sudden infant death syndrome	Infants <1 year	Postnatal maternal smoking	

LRI- lower respiratory tract infection, LBW-low birth weight, SGA-small for gestational age

Indoor inorganic pollutants

Indoor inorganic contaminants include CO_2, CO, SO_2, NO_2, NO_2, ozone (O_3), hydrogen chloride (HCl), nitrous acid (NHO_2), nitric acid vapour (HNO_3) and radon. When no combustion source is present, CO_2 level inside an occupied building should be no more than 650 ppm above outdoor levels. Mainly household combustion generates elevated indoor levels of CO_2, SO_2 and NO_2. It has been reported (epidemiological studies) long term exposure to NO_2 through use of gas stove is a modest risk factor for respiratory illness (2) (Basu and Samet, 1999). On the other hand CO causes acute poisoning i.e. it is able to bind haemoglobins strongly and is a leading cause of poisoning death in USA and claims many lives worldwide (Girman et al., 1998). Indoor ozone, penetrating from outdoors or derived indoors can drive chemical reactions among chemical species present indoors, generating secondary pollutants which cause various health problems (Weschler, 2001). Source of HCl include outdoor to indoor transport and thermal decomposition of polyvinyl chloride (PVC) and is a real health concern. Major source of indoor HNO_3 include penetration of outdoor HNO_3 formed in photochemical smog and HNO_3 formed indoors via reactions include O_3, NO_2, and water vapor (Weschler et al., 1992). Acidic gases e.g. HCl and HNO_3 are highly corrosive to materials and biological tissues.

Indoor Organic Pollutants

Indoor organic contaminants are characterized by volatility. VOCs (volatile compounds) present in the gas phase at typical indoor concentrations and have boiling point of from <0°C to 240-260°C. SVOCs (semi-volatile organic compounds) are the partitioning between the gas phase and the particulate phase under typical indoor conditions and have boiling points from 240-260 to 380-400°C. Indoor organic compounds released from a variety of building materials such as vinyl tile and covings. Some VOCs and SVOCs are mutagenic and/or carcinogenic e.g. benzene, styrene, tetrachloroethylene, 1,1,1-trichloroethane, trichloroethylene, dichlorobenzene, methylene chloride and chloroform. Long term exposure to these compounds produces cancer risks. Many VOCs and SVOCs e.g. aldehydes have the potential to cause sensory irritation and central nervous system symptoms (pesticides). It has been reported that paternal exposure to VOCs e.g. chlorinated solvents, spray paints, dyes and pigments, cutting oils during work and maternal VOCs exposure during pregnancy are responsible for increased risk of childhood leukemia (Godish, 2000).

Effect of Indoor Pollution on Human health

The impact of indoor air pollution on man may consist of undesired health effects of different types, ranging from sensory annoyance or discomfort to severe health injuries. "Health" is defined, for the purpose of this report, according to the well-known WHO definition as "A state of complete physical, mental and social well-being, and not merely the absence of disease or infirmity". The public health relevance of the effects of IAP varies, not only from substance to substance, but also from country to country, depending on the presence of specific local sources and climatic influences.

General matrix for the evaluation of the impact of indoor pollutants on the community:

The "Sick Building syndrome"(SBS)

Since the early 1970s, numerous outbreaks of work related health problems have been described among employees in buildings or offices not directly contaminated by industrial processes. Two broad categories can be distinguished: those characterized by a generally uniform clinical picture for which a specific cause has been identified, and those in which affected workers reported nonspecific symptoms occurring only during the time when they were at work. The former episodes have been defined "Building - Related Illness" (BRI), the latter, "Sick Building Syndrome" (SBS). Symptoms reported in SBS have typically included mucous membrane and eye irritation, cough, chest tightness, fatigue, headache and malaise. Outbreaks without an identifiable cause have frequently occurred in new, sealed office buildings and have for that reason also been called the "tight building syndrome" (TBS).

Symptoms and reactions observed belong to the following groups

A. Acute physiological or sensory reactions
- Sensory irritation of mucous membranes or skin
- General malaise, headache and reduced performance
- Unspecific hypersensitivity reactions, dryness of skin
- Odor or taste complaints

B. Psychosocial reactions
- Decreased productivity, absenteeism
- In contact with primary health care
- Initiatives to modify the indoor environment
- Sensory irritation in eyes, nose and throat must be dominating
- Systemic symptoms (e-g. from stomach) must be infrequent
- No obvious causality can be identified in relation to high exposure to single agents.

Non Specific building related Illness (NSBRI)

NSBRI characterized by health symptoms e.g. mucous membrane irritation (ocular, nasal), headache, fatigue, shortness of breath, rash and odor complaints (Fiedler et al., 2002) Studies reported that indoor pollutant mixtures along with physicological mixtures may play an important role in causing NSBRI. It has been demonstrated that in some cases, VOC exposure associated with neurobehavioral performances (Molhave et al., 1986). lung functions (Kjaergaard et al., 1991) and nasal inflammation (Koren and Devlin, 1992).

Prevention and Control of IAQ Problems

IAQ Management Approach

Ideally, an employer should use a systematic approach when addressing air quality in the workplace. The components of a systematic approach for addressing IAQ are the same as those for an overall safety and health program approach, and include management commitment, training, employee involvement, hazard identification and control, and program audit. Management needs to be receptive to potential concerns and complaints, and to train workers on how to identify and report air quality concerns. If employees express concerns, prompt and effective assessment and corrective action is the responsibility of management. It is recommended that building owners/managers develop and implement an IAQ management plan to address, prevent, and resolve IAQ problems in their specific buildings. The EPA's report, *IAQ Tools for Office Buildings*, provides a set of flexible and specific activities that can be useful to building owners/managers for developing such a plan. A key feature of the plan is the selection of an IAQ Coordinator. The role and functions of an IAQ Coordinator are described in Section 3 of the EPA's report, *IAQ Tools for Schools Action Kit* (CDC, 1990). Other critical features of the plan include establishing necessary IAQ policies, assessing the current status of IAQ in buildings through periodic inspections, maintaining appropriate logs and checklists, performing necessary repairs and upgrades, and implementing follow-up assessments or other needed actions. Employers who lease space should be familiar with the building management's program and methods for mitigating or resolving indoor air quality problems. It is especially important for employers to know who to contact in buildings where there is mixed use and pollutants are emanating from other sources in the building. Employers should negotiate leases that specify IAQ performance criteria. An important management strategy is to foster a team approach for problem solving and consensus building. The IAQ Team should include, but not necessarily be limited to, building occupants, administrative staff, facility operators, custodians, building healthcare staff, contract service providers, and other interested parties. Lastly, following up with affected personnel will serve to validate the effectiveness of the mitigation activities. For more information about the IAQ management approach, refer to OSHA's Safety and Health Topics Page on Injury and Illness Prevention Programs. (OSHA, 2011).

REFERENCES

1. Alcaide J, Altet MN, Plans P, et al., Cigarette smoking as a risk factor for tuberculosis in young adults: A case control study. Tubercle Lung Dis 1996; 77:112-16;
2. Altet MN, Alcaide J et al., Passive smoking and risk of pulmonary tuberculosis in children immediately following infection. Acase control study. Tubercle Lung Dis 1996; 77:537-544.
3. Apte MG, Fisk, WJ, Daisey JM Association between indoor CO_2 concentration and sick bulding syndrome symptom in U.S. office buildings: an analysis of the 1994-1996 BASE study area. Indoor air 2000; 10; 246-57.
4. Basu, R, Samet, JM. A Review of the epidemiological evidence on health effects of nitrogen dioxide exposure from gas stoves. J Environmental Med 1999; 1: 173-7.
5. Center for Disease control. The Surgeon General's 1990 Report on Health Benefits of Smoking Cessation Executive Summary. MMWR 1990; 39: viii-xv.

6. Fiedler N, Zhang J, Fan Z et al., Health effects of a volatile inorganic mixture with and without ozone. Proceeding of indoor air 2002.

7. Girman JR, Chang, YL, Hayward SB, Liu KS. Causes of unintentional deaths from carbon monoxide poisoning in California. West J Med 1998; 168: 158-65.

8. Godish T., Organic contaminants. In: Indoor Environmental quality. Boca Raton, FL: CRC press 2000; 95-141.

9. Khuder, S.A. and Simon Jr VJ. Is there is an association between passive smoking and breast cancer? Eur J Epidemiol 2000; 16:1117-21.

10. Kjaergaard SK, Molhave L, Pederson OF. Human reactions to a mixture of indoor air volatile organic compounds. Atmos Environ 1991; 25 A: 1417-26.

11. Koren HS and Devlin RB. Human upper respiratory tract responses to inhaled pollutants with emphasis on nasal lavage. Ann N Y acad Sci 1992; 641: 215-24.

12. Molhave L, Bach B, Pedersen OF. Human reactions to low concentrations of volatile organic compounds. Environ Int 1986; 12: 167-75)

13. Occupational Safety and HealthAdministration U.S. Department of Labor OSHA Indoor Air Quality in Commercial and Institutional Buildings. 3430-04. 2011.

14. Smith, KR, Indoor air pollution in developing countries: recommendations for research. Indoor air 2002; 12: 198-207.

15. Tsia et al., 2003. Tsia, SMnZhang, J., Smith, KR, Ma Y, Rasmussen, RA Khalil MAK. Characterization of non methane hydrocarbons emitted from various cookestoves used in china. Environ Science Technol, 2003; 37: 2689-2877).

16. U.S. Environmental Protection Agency (EPA). Respiratory health effects of passive smoking: Lung cancer and other disorders. Washington DC: US Environmental Protection Agency, 1992; EPA/600/6-90/006F (NTIS PB 93-134419).

17. Weschler CJ reactions among indoor air pollutants. Sci World 2001; 1:443-57).

18. Weschler CJ, Brauer M, Koutrakis P, Indoor ozone and nitrogen dioxide: A potential pathway to the generation of nitrate radicles, dinitrogen pentoxide, and nitric acid indoors. Environ Sci Technol 1992; 26: 179-84).

19. WHO World health report: Reducing Risks, Promoting healthy Life. Geneva: WHO, 2002.

20. Zhang, J., and Smith, K.R. (2003). Indoor air Pollution: A global health concern. British Medical Bulletin, 68: 209-225.

21. Zhang, J., Smith KR, Ma, Y. et al. Greenhouse gases and other pollutants from household stoves in China: A database for emission factors. Atmos Environ 2000; 34: 4537-49.

22. Zhang, J. Smith KR. Emission of carbonyl compounds from various cookstoves in china. Environment Science Technol 1999; 33: 2311-20.

23. Zhang, J., Smith, KR, Hydrocarbon emissions and health risks from cookestoves in developing countries. J Expose Anal Environ 1996; 33: 2311-20.

ISBN-13: 978-1533168078
ISBN-10: 1533168075

Chemistry of Industrial Globalization, Environmental Pollution and its Chem-Biological Significance

February 2016

Industrial Globalization and Environmental Awareness
Proceedings of the NATIONAL SEMINAR held by
Department of Chemistry, Government PG College, Ambala Cantt - Haryana - India

A Survey of Encryption Techniques to Secure Cloud Storage

Himshikha Rahi

UCOE, Punjabi University
Patiala, India
hshikha46@gmail.com

Abstract

Cloud computing is an emerging technology which uses a network of remote servers hosted on the Internet to store, manage, and process data. Cloud computing provides easy and user friendly platform to a company's high-performance computing and storage infrastructure through web services. The services are offered from data centers all over the world known as "cloud". Data is stored in various data centers i.e. in a third party's cloud system which causes serious concern on data confidentiality. In order to provide confidentiality and security to the data stored in cloud, cryptography method is used. It is a method of encrypting plain text to cipher text using some encryption techniques. The encryption is done at different levels such as when data is at rest and when data is in motion. This paper briefly reviews about the few encryption techniques used to secure encryption. The encryption techniques protect data confidentiality but also limit the functionality of the storage system because few operations are supported over encrypted data.

Keywords- Cloud Computing, Cloud Security, Encryption, Cloud Storage

1. Introduction

Cloud Computing is a type of computing which deals with sharing computing resources instead of using local servers or personal devices to handle applications. In cloud computing, the word "cloud" is used as a metaphor for "the Internet". So the cloud computing is a type of internet based computing in which different servers such as servers, storage and applications are made available to the organization's computers and devices through internet.[1]

In cloud storage system, data is stored in data centers which cannot be trusted. So, user would like to check whether the data has been tampered or deleted. In order to protect cloud computing, no new technique is required. Protecting data in the cloud is similar to protecting data in the traditional Data centers. There are so many protection methods such as Authentication and identity,

Access control, encryption, secure deletion, integrity checking, and data masking. [2] This paper briefly reviews about the few encryption techniques used to secure encryption.

Characteristics of Cloud Computing

- **Application Programming Interface (API):** It acts as an interface which provides accessibility to software which enables machine to interact with the cloud in a way humans interact with the computers in a traditional user interface e.g. computer desktop.
- **Device and Location Independence:** Cloud computing enables user to directly access the cloud using some web browser with the help of internet regardless of the location and device.[3]
- **On Demand Self Services:** Cloud service providers provide various on demand self services without requiring the human interactions such as email, application, network or server.
- **Broad Network Access:** The capabilities of the cloud are available over the network and can be accessed through various standard mechanisms such as thick or thin client platforms such as mobile phones, laptops etc.[4]
- **Productivity:** Productivity in cloud computing can be increased as multiple users can work simultaneously. Also users do not have to install application software upgrades to their computers.[3]

2. Service Models

According to NIST, the cloud model is composed of three service models:

o **Software as a service (SaaS):**
The software is installed directly on the cloud; there is no need of installing the software individually on local personal computers or local servers. Cloud providers install the application software on the cloud itself which is then accessed by the cloud clients. It is also referred to as 'on-demand' service. E.g. Blogs, E-mail, newsletter, PDF conversion, etc.

o **Platform as a service (PaaS):**
Application developers can develop and run their software solutions on a cloud platform without the cost and complexity of buying and managing the underlying hardware and software layers. Services that are provided as a platform includes operating systems, programming language execution environment, database and web server. E.g. Oracle, Google apps, Windows Azure.

o **Infrastructure as a service (IaaS):**
In this model, physical resources are abstracted by virtualization which means they can be shared by several operating systems and end user environments on the virtual resources without any mutual interference. IaaS cloud providers supply resources such as virtual machine disk image library, raw block storage, and file or object

storage, firewalls, load balancers, IP addresses, virtual local area networks, etc on demand from their large pools installed in data centers. E.g. ASP.net, PHP, Open stack, Amazon web services, etc. [3]

3. Encryption

The cloud service providers must ensure that their infrastructure is secure and that their client's data and applications are protected whereas the user must ensure that the provider has taken the proper security measures to protect their information and the user must take measures to use strong passwords and authentication measures to secure their data in the cloud. So the security issues are faced by cloud providers (organizations providing software, platform or infrastructure as a service via the cloud) and their customers. For protection of data in cloud there is no new protection technique is required. Protection of data in the cloud is similar to the protection of data within a traditional data center. Various data protection methods are used in cloud computing such as authentication, and identity, access control, encryption, secure deletion, integrity checking and data masking. This paper will briefly review some of the encryption techniques which can be used to protect data in the cloud storage. [2]

In cloud computing, it is frequent for the entities to communicate manually to achieve the security in the communication, it is important to impose an encryption technique. With encryption we ensure that both, the data in motion and at rest, are completely secured through encryption. Cloud encryption is a service offered by cloud storage providers whereby data or text is transformed from plain text to cipher text through some encryption algorithms and is then placed on a storage cloud. The goal of encryption is to ensure that data stored in the cloud is protected against unauthorized access `encrypting data in transit also helps to secure it as it is often difficult to physically secure all access to networks.[5]

Encryption is a key component to protect data at rest in the cloud. Strong encryption is preferable when data at rest has continuing value for extended time period. If such long term value encrypted data is obtained by a third party and if they have an extensive period of time to break or crack the encryption, then the reward can be well worth the effort. [6]

Application of Encryption for data at rest:

There are multiple ways of encrypting data at rest. Following are some of the ways to encrypt data at rest in the cloud:

- ❖ **Full disk:** In this level the operating system, applications in it and data stored in it are encrypted as well as the storage disk is also encrypted.
- ❖ **Directory Level:** In this a directory or hard drive is encrypted or decrypted as a container. And when we need to access the files we use encryption keys. In this directories are categorized into different categories which can be accessed by using different encryption keys.
  - ➢ **File level:** Instead of encrypting the whole drive or the directory we can encrypt a particular file.
  - ➢ **Application Level:** Encryption and decryption of application managed data is done at this level.[2]

Application of Encryption for data in motion:

There are two major goals of securing data in motion:
- preventing data from being tampered that is preventing its integrity,
- ensuring that data remains confidential.

- ❖ **Data Encryption Standard (DES):** It is a block cipher. It encrypts data in blocks of 64 bits each. 64 bits of plain texts goes as the input to the DES, which produces 64 bits of cipher text. The key used for the processes 64 bit from which every 8th bit is discarded to form a 56 bit key.DES is based on two fundamental attributes of cryptography: substitution (also called confusion) and transposition (also called diffusion).
- ❖ **Advanced Encryption Standard (AES):** The 56 bit key was no longer considered safe against attacks and also the 64 bit blocks were also considered weak. So the US govt. standardized a cryptographic algorithm which can be used universally. AES is based on encrypting 128 bit blocks with 128 bit keys.
- ❖ **International Data Encryption Algorithm (IDEA):** It is considered as the strongest cryptographic algorithm. IDEA is a block cipher. It uses 64 bit plain text blocks and produces 64 bit cipher text with the help of 128 bit key. IDEA uses substitution and transposition techniques for encryption. In this, the 64 bit input plain text is divided into 4 portions of plain text each of size 16 bit. The key is also divided into 6 portions each of size 16 bit. It has 8 rounds which further produces 64 bit cipher text block.
- ❖ **RSA Algorithm:** It is one of the first practicable public key crypto-systems and is widely used for secure data transmission. In such a crypto-system, the encryption key is public and differs from the decryption key which is kept

secret. In RSA, this asymmetry is based on the practical difficulty of factoring the product of two large prime numbers i.e. the factoring problem.[7]

❖ **Searchable Encryption:** It provides a way to encrypt a search index so that contents are hidden except to a party that is given appropriate tokens. Using a searchable Encryption scheme, the index is Encrypted in such a way that given a token for a keyword one can retrieve pointers to the encrypted files that contains the keyword and without a token the contents of the token are hidden.[6]

❖ **Proxy Re- Encryption:** It transforms cipher text ck1 to cipher text ck2 with a key rkk1->k2 without revealing the corresponding clear text. Ck1 and ck2 can only be decrypted by different key k1 and k2 respectively and rkk1->k2 is a re key issued by another party, e.g., the originator of cipher text ck1.[8]

❖ **Identity based Encryption for Hierarchical architecture for Cloud Computing (HACC):** Identity based systems allow any party to generate a public key from a known identity value. A trusted third party, called the private key Generator (PKG), generates the corresponding private keys. The PKG first publishes a master public key, and retains the corresponding master private key. Given the master public key, any party can compute a public key corresponding to the identity ID by combining the master public key with the identity value. To obtain a corresponding private key, the party authorized to use the identity ID contacts the PKG, which uses the master private key to generate the private key for identity ID.[9]

❖ **Key Policy Attribute-Based Encryption (KP-ABE):** KP-ABE is a public key cryptography primitive for one to many communications. It enables to encrypt messages under a set of attributes and private keys are associated with access structures that specify which cipher texts the key holder will be allowed to decrypt. A KP-ABE scheme is composed of four algorithms which are setup, Encryption, Key Generation, and Decryption.[10]

3. References

1. http://www.webopedia.com/TERM/C/cloud_computing.html
2. Satyendra Singh Rawat and Mr. Alpesh Soni, A survey of various techniques to secure cloud storage, National Conference on Security Issues in Network Technologies (NCSI-2012) August 11-12,2012
3. http://en.wikipedia.org/wiki/Cloud_computing

4. http://www.isaca.org/groups/professional-english/cloud-Computing/groupdocuments/essential%20characteristics%20of%20cloud%20computing.pdf
5. http://en.wikipedia.org/wiki/Encryption
6. Mahima Joshi and Yudhveer Singh Moudgil, Secure Cloud Storage, Mahima Joshi et al, International Journal of Computer Science & Communication Networks,Vol 1(2), 171-175, ISSN:2249-5789
7. Atul Kahate, Cryptography and Network Security
8. S.Poonkodi, V.Kavitha, K.Suresh , Providing A Secure Data Forwarding In Cloud Storage System Using Threshold Proxy Re-Encryption Scheme
9. http://www.hindawi.com/journals/mpe/2013/810969/
10. http://en.wikipedia.org/wiki/Attribute-based_encryption

ISBN-13: 978-1533168078
ISBN-10: 1533168075

Chemistry of Industrial Globalization, Environmental Pollution and its Chem-Biological Significance
February 2016

Industrial Globalization and Environmental Awareness
Proceedings of the NATIONAL SEMINAR held by
Department of Chemistry, Government PG College, Ambala Cantt - Haryana - India

Green IT - Go for Green Gadgets

Veenu Saini

Assistant Professor (IT), Post Graduate Govt. College-11, Chandigarh
Ph.D. Research Scholar (CSE), PTU, Jalandhar
vnusaini@gmail.com

Abstract

The goal of any technology is to make our lives better. We live in an Information Technology (IT) era where computers, data centers, servers, internet and other machineries are used in almost all industries. From a sustainability viewpoint, however, the high usage of technology brings out several issues, such as extreme power consumption and an expanding carbon footprint that has a detrimental effect on the environment. Increasingly, these issues induce a demand for environmental sustainability concerns because of global societal concerns about energy usage, climate change, and the consequences of these changes. These concerns are important to the IT sectors because high usage of technology does contribute to the increase of greenhouse gas emissions. In response to these challenges, the term Green IT has been coined with reference to initiatives that are focused on reshaping IT into environmentally friendly forms.

'Green IT' addresses the direct impact of energy consumption and waste associated with the use of hardware and software. The green information technology movement and spawned development of hundreds of apps, websites, companies and organizations dedicated to producing new technology solutions with sustainability in mind.

IT products such as smart phones, computer tablets, music players, video players, and gadgets increasingly come is various shades of Green – designed to use less electricity and produce less hazardous waste, manufactured using more recycled materials and less energy. This paper focuses on using the eco friendly products.

Keywords Green technology, Green Gadgets, Eco friendly Products, Green IT

Introduction

Green IT promotes the design and manufacturing and management of IT equipment and services that consume minimal energy throughout their entire life-cycle (Murugesan 2008). Therefore, it can be argued that Green IT refers to the use of IT resources in an energy-efficient as well as cost-effective manner. The term 'Green IT' also spurs the hope that IT will be an effective means for organizations looking to make significant steps in the reduction of the environmental impact of their operations.

Some of the Eco friendly Green IT Gadgets are shown below and discussed in material, method and discussion section.

Literature Search

Research on Green Information Technology (IT) is becoming a prevalent research theme in Green Information Systems (IS) research. I surveyed through many literatures which facilitate me for future research and to provide a retrospective analysis of existing knowledge and gaps thereof. While some researchers have discussed phenomena such as Green IT, motivation of Green IT and the Green IT adoption lifecycle, others have researched the importance of Green IT implementation within the organisational and individual level. Throughout the literature, scholars are trying to portray a constructive relationship between IT and the environment. In this paper my main emphasis is on the use of Eco friendly products such as A Bright Laptop Case, Genius Mouse, Flash Drive, Solar Powered Keyboard, The Nest, Web Hosting, Desktop addition, Phone Charger, iBamboo, iZen Bamboo Bluetooth Keyboard, SteriPEN Ultra, USBCell Rechargeable Batteries and Eco-Amp etc.

Materials, Methods and Discussion

Brief about Green Gadgets

The Nest Google made big waves when it bought thermostat company Nest for $3.2 billion. The "next generation thermostat" goes green at home with the Internet of Things by automatically adjusting home heat. This save up to 20 percent on monthy bill.	**Genius Mouse** Claiming to be the "most eco-friendly wireless mouse," the Genius DX-ECO 2.4 GHz BlueEye wireless mouse is both comfortable to use and good for the environment. The mouse is completely battery-free. The mouse takes only three minutes to charge before it is ready to go with its wireless technology.	**Web Hosting** Data center company AISO.net are powered by on-site solar panels, water evaporation cooling systems that obtain water through rainfall and use atmospheric energy for air conditioning. The offices use mirrored tubes and windows to light the building, and they are covered with a "green roof" to cool the building.

Phone Charger Mushroom GreenZero wall charger is the most eco-friendly way to charge a phone. When other chargers are left plugged in, they still pull power even when they aren't charging a device. This Charger eliminates that unused power, lowering energy bills and cutting back on waste.	**Flash Drive** This SanDisk flash drive takes files on-the-go in eco-friendly style. The 16-GB USB flash drive is encased in bamboo rather than plastic or metal, making it more eco-friendly because bamboo grows quickly and releases more oxygen than other trees.	**Solar-Powered Keyboard** This Logitech keyboard works completely off solar power, which means it needs neither batteries nor a plug to be up and running, even indoors and without any cables. The Logitech Solar App can be downloaded right to a mobile device to keep track of battery power.
		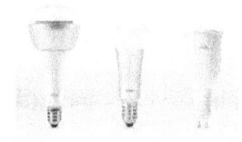
iBamboo iBamboo portable speakers are bamboo cut to the usual portable speaker size with a slot on the top to rest your iPhone. Turn your phone's speaker on, set it in the slot, and the natural acoustics of the bamboo will amplify the sound, no electricity required. iBamboo is compatible with iPhone 4 through 5s and is available in natural and black.	**Eco-Amp** It amplifies sound and is made in the U.S. from renewable materials. The Eco-Amp is a passive speaker and increases sound volume and clarity without external power. It was originally designed for iPads and iPhones, but now works on nearly all devices.. It costs $10.00 for both phones and tablets.	**Philips Hue Connected Bulb** The Philips bulbs connect over your Wi-Fi, allowing you to use an app to control hue and dimming for each individual bulb. They work like regular bulbs and they still use less energy.

A Bright Laptop Case	iZen Bamboo Bluetooth Keyboard	Eton Mobius
Although it isn't green in color, this laptop case's material is 100 percent eco-friendly. The exclusive felt-tech material is made completely from recycled bottles.	Bamboo grows five times as fast as wood and can be grown in a variety of environments, which is what makes it sustainable and the go-to, eco-friendly material nowadays	Eton put out a rechargeable battery for the iPhone 4 and 4s that is essentially a case with a solar panel attached to the back. Called Mobius,

USBCell Rechargeable Batteries	Water Pebble	SteriPEN Ultra
USBCell offers AA batteries that operate and look just like normal AA, but the positive end opens to reveal a USB plug that will connect to any USB charger, including your computer.	Water Pebble helps you reduce your shower time in a simple way. Put it near the drain and it will monitor how much water goes down. it will display lights—green, orange, and red.	The SteriPEN Ultra eliminates 99.9 percent of bacteria from water in just 48 seconds. The UV lamp can be used to treat 8,000 one-liter containers.

Conclusion

Green IT has already had a measurably positive effect on efforts to reduce energy consumption since its recommendations have been applied. Another prominent theme featured in these analyses is a recurring emphasis on the various ways that Green IT has been responsible for detrimental outcomes resulting from the overuse of information networks. Green technologies have upsides and downsides but they are a necessary approach towards human survival. In the long run they have been proven to be beneficial to the society but their true effects can be observed only in the future which we can safely hope to be good for the society.

Significance

A significant amount of research has been performed on investigating Green IT initiatives. The initiatives range from technical solutions for more sustainable IT to a soft solution approach to promote Green IT. The technical solution focuses on the energy-efficient equipment and eco-friendly hardware in terms of using, designing and manufacturing, and disposing (Murugesan 2008).

On the other side, soft solutions draw the attention of behavioural attitudes like paperless offices, less printing or printing on both sides (Ansari et al. 2010), video and mobile conferencing. Apart from explaining environmentally friendly hardware, many scholars propose soft solutions to promote Green IT. Vlek and Steg (2007) identify many challenges of human behaviours and environmental sustainability, including the significant effect of behavioural and/or environmental changes on human well-being.

Bibliography

1. Ansari, N.L., Ashraf, M.M., Malik, B.T., and Grunfeld, H. 2010. "Green It Awareness and Practices: Results from a Field Study on Mobile Phone Related E-Waste in Bangladesh," International Symposium on Technology and Society (ISTAS) pp. 375-383.
2. Murugesan, S. 2008. "Harnessing Green It: Principles and Practices," IT Professional (10:1), pp. 24-33.
3. Vlek, C., and Steg, L. 2007. "⊡ Human Behavior and Environmental Sustainability: Problems, Driving Forces, and Research Topics," Journal of social issues (63:1), pp. 1-19.
4. www.envirogadgets.com/
5. https://en.wikipedia.org/wiki/Environmentally_friendly

ISBN-13: 978-1533168078
ISBN-10: 1533168075

Chemistry of Industrial Globalization, Environmental Pollution and its Chem-Biological Significance February 2016

Industrial Globalization and Environmental Awareness

Proceedings of the NATIONAL SEMINAR held by

Department of Chemistry, Government PG College, Ambala Cantt - Haryana - India

The Effects and Implications of Climate Change on Plant Phenology

Dr. Rohini Singh[1] and Dr. Sheesh.P. Singh[2]

[1]Asst. Prof. & H.O.D. Deptt. of Botany, Govt. (PG) College, Ambala Cantt.

[2]Asst. Prof. Deptt. of Botany J.V. (PG) College Baraut (Baghpat)

rohishum@gmail.com[1]   Sheesh9@yahoo.in[2]

ABSTRACT

Global climate change could significantly alter plant phenology because temperature influences the timing of plant development. Plant phenology is finely tuned to the seasonality of their environment and any shift in plant phenology can be taken as evidence that species and ecosystems are rescheduling their activities in response to global climate change. Plants and animals exhibit seasonal patterns in their activities because there is a clear seasonality in the suitability of their environment: there is often only a limited period in the year when conditions are favourable enough to successfully reproduce or grow. If reproduction or growth takes place in sub - optimal conditions, there are often variable fitness consequences. Ultimately, the activity that is the most demanding for an organism such as flowering and pollination in plants should take place under the optimal conditions. Experimental and modeling approaches have sought to identify the mechanisms causing these shifts, as well as to make predictions regarding the consequences. The observed changes in phenology cannot be interpreted without considering the ecological context in which a species lives, and especially how other components of the ecosystem are affected by climate change. Global change experiments have also documented the influence of increasing temperature, shifting precipitation and other aspects of global change, such as rising CO2 concentrations on the timing of species and ecosystem level phenology. Accurate phenology predictions would increase the accuracy of the predictions of ecosystems productivity and gas exchanges with the atmosphere, better understanding of population dynamics in multi-species interactions systems. This could help in better management of agriculture and forests.

Keywords: Climate change, Plant phenology, temperature ranges, vegetative and reproductive phases, leaf unfolding and flowering, El Niño, tropical ecosystems, ecosystems productivity, gas exchanges, agriculture, satellite images, phenology predictions.

Introduction

Climate change has been recognized as one of the greatest threat to life on Earth. The changes in seasonal patterns, weather events, temperature ranges, and other meteorological phenomena have been reported and attributed to climate change. Scientists have warned that the negative impacts of climate change will become more intense and frequent in the future, unless environmentally destructive human activities continue unabated.

Plant phenology is finely tuned to the seasonality of their environment. Any shift in plant phenology can be taken as evidence that species and ecosystems are rescheduling their activities in response to global climate change. Scientists have observed shifting phenology at multiple scales, including earlier spring flowering in individual plants and an earlier spring green-up of the land surface viewable through satellite images. Experimental and modeling approaches have sought to identify the mechanisms causing these shifts, as well as to make predictions regarding the consequences. The observed changes in phenology cannot be interpreted without considering the ecological context in which a species lives, and especially how other components of the ecosystem are affected by climate change. Plants and animals exhibit seasonal patterns in their activities because there is a clear seasonality in the suitability of their environment: there is often only a limited period in the year when conditions are favourable enough to successfully reproduce or grow. If reproduction or growth takes place in sub - optimal conditions, there are often variable fitness consequences. Ultimately, the activity that is the most demanding for an organism (flowering and pollination in plants) should take place under the optimal conditions. What we need to know to evaluate the observed shifts in phenology is how that period of optimal conditions shifts due to climate change. The shift in the seasonal changes in the ecological conditions could be used as a yardstick to assess whether the change in phenology observed is sufficient or not.

Phenology is a dominant and often overlooked aspect of plant ecology, from the scale of individuals to whole ecosystems. The timing of the switch between vegetative and reproductive phases that occurs in concert with flowering is crucial to optimal seed set for individuals and populations. The variation among species in their phenology is an important mechanism for maintaining species coexistence in diverse plant communities, by reducing competition for pollinators and other resources. The timing of growth onset and senescence also determine growing season length, thus driving annual carbon uptake in terrestrial eco-systems. Plant phenology is strongly controlled by climate and has consequently become one of the most reliable bio-indicators of ongoing climate change. Phenology has received much attention during the last decade because many organisms are changing their life cycles in response to ongoing climate change (Parmesan & Yohe, 2003; Menzel et al., 2006a; Rosenzweig et al., 2008). Plants are especially appropriate organisms to study climate effects in phenology because they are sessile and thus must endure all weather conditions occurring where they are located. Such sessile life-style has led plants to show an especially high plasticity in their phenotypes, such as phenology, to deal with environmental variability (Schlichting, 1986). For instance, Fitter et al. (1995) found that flowering timing in 90% of 243 studied plant species in England was significantly related to temperatures, i.e. the overwhelming majority of plants were able to tune their flowering dates according to particular temperature conditions of each year. This strong dependence on climate explains why, of all taxonomic groups, plants have the highest portion of species shifting their phenology in the predicted direction under current climate change (Parmesan & Yohe, 2003). Temperature has been demonstrated as one of the most important factors for plant phenology (Sparks et al., 2000; Pen~uelas et al., 2002; Matsumoto et al., 2003; Menzel, 2003; Gordo & Sanz, 2005; Ahas & Aasa, 2006; Estrella & Menzel, 2006; Lu et al., 2006; Menzel et al., 2006a).

Methods

Phenological networks rely on volunteers to collect observations of various phenol-phases of wild plants, fruit trees and agricultural crops at numerous stations. The longest and best known phenological records come from the Far East and Europe, including a 5000-year record of phenological events, weather and farming activities in China , the 1300+year Kyoto cherry blossom time series , 670+ years of grape harvest dates in Central Europe , and the 200+year Marsham record of plant and avian phenology in the UK. These long term historical records can serve as proxies for temperature where thermometer data are unavailable. More recently, shifting phenology from the mid-20th century onwards is evident from numerous phenological observation networks in the Far East, North America and Europe. A meta-database of existing networks shows that most observation networks are located in temperate ecosystems and that long-term phenological observations in the tropics are lacking

Two recent meta-analyses of observational data have found that spring has advanced globally at a rate of 2.3 days per decade and 5.1 days per decade, respectively. The second study excluded records exhibiting no trend. The most comprehensive analysis of observed trends in phenological data has been carried out for Europe. This study included >125 000 observational series of 542 plant and 19 animal species in 21 European countries (1971– 2000) and found a coherent signal of earlier spring (leaf unfolding and flowering) and summer (fruit ripening) of 2.5and 2.4 days per decade, respectively. More than 75% of the species had accelerated phenology over this time period (the acceleration was statistically significant for one third of the total species). Variation among species in their phenological trends has been attributed to pollination type, life form and time of year. Early-season species exhibit the greatest acceleration, possibly because their phenology is cued by melting snowpack. Interestingly, farmers' decisions about which crops to plant in a given year have resulted in the phenology of agricultural species changing less than those in the wild and, as a consequence, farmers' activities in Europe have only advanced by 0.4 days per decade. The ground-observational signal of leaf coloring and leaf fall in autumn is more variable than is leaf unfolding in spring; some areas show acceleration whereas others show delay. By contrast, remotely sensed and phenological-climatic measures consistently find that the onset of autumn has been delayed in recent years. Ground-observation data averaged across continental Europe suggest a delayed autumn of 1.3 days per decade, whereas results for some regions within Europe show little change.

There are two major approaches in phenological studies; one is the individual scientific studies such as species-specific phenological observations that are able to scale from local to regional shifts in phenology as well as to predict shifts in species ranges in response to climate change. The other approach is the Satellite observations to study phenological and ecological responses to environmental changes over space and time. Although specific phenological-phases such as flowering and fruiting cannot be discerned in satellite images, their combination of comprehensive ground coverage and regularly repeated observations offer the opportunity for global phenological monitoring that is not possible with any other source.

Analyses to detect temporal trends in phenology have traditionally relied on regressions or correlations between observed events and variables, such as year or temperature and, less frequently, have utilized time-series analysis. These traditional methods cannot detect abrupt phenological changes that can accompany rapid shifts in climate; however, new techniques based on change point analysis, and Bayesian techniques enable a quantitative representation of non-linear phenological responses and associated rates of change.

Discussion

Global climate change could significantly alter plant phenology because temperature influences the timing of development, both alone and through interactions with other factors, such as photoperiod the records show that, over the past 30 years, global average surface temperatures increased by $0.28^{0}C$ per decade.

In addition, numerous studies examining frost dates, growing season length, growing degree totals have found changes that are consistent with climate warming. These phenoclimatic measures represent changes in temperature that are relevant for different phases of plant development. The phenoclimatic measures, found that first leaf dates and last frost dates were 1.2 days and 1.5 per decade earlier, respectively, for Northern Hemisphere temperate land areas from1955 to 2002 .This acceleration of spring has led to a longer growing season, predominantly via warming of the coldest days in late winter and early spring as opposed to consistent warming throughout the year. These phenoclimatic measures have important applications for agriculture, as well as for parameterizing leaf phenology in global climate models

In tropical ecosystems, phenology might be less sensitive to temperature and photoperiod, and more tuned to seasonal shifts in precipitation. Such shifts are expected to occur in concert with rising global temperatures, but both the direction and magnitude of change vary regionally. For areas where precipitation patterns are strongly influenced by the El Niño -Southern Oscillation, the frequency and intensity of El Niño events are expected to increase, which has already begun to affect tropical forest phenology in Brazil

In addition to shifting phenology, species have begun to adapt to recent climatic changes via altered species ranges. It will be important to address whether the combination of physiological shifts within species and range shifts among species will alter community-level patterns of phenology. Species within communities are often remarkably varied in their phenology, and we are just beginning to understand how potential shifts in phenological complementary might feedback to influence ecosystem structure and function. Aerobiology, or the study of pollen composition, is a promising new approach for studying the phenology of whole communities, although it is taxonomically limited in scope (dominated by trees) and resolution (identified at the genus or family level), and there is uncertainty around the degree of long-distance pollen transport . Shifts in the synchrony of plant–animal interactions could also negatively impact particular plant populations, and re-order communities. The best studies of disrupted synchrony involve trophic interactions. For instance, in some migrating birds, the timing of egg hatching has become asynchronous with the availability of insects that are an essential food source for hatchlings and, in marine ecosystems, the timing of phytoplankton availability has cascaded up the food chain to impact fish predators. These observations have been largely attributed to rising temperatures over the same time period, but a review of experimental studies shows that elevated CO_2 can also change the phenology of plants and insect herbivores in opposing directions. The study of synchrony between plant and animal mutualisms is an area of much needed research. For instance, if pollinators were cued to temperature and plants to photoperiod, this could lead to asynchrony in pollinator mutualisms.

However, its true relevance for plants could be overestimated, since few studies have assessed the effect of other environmental factors such as precipitation, photoperiod, and availability of soil nutrients or soil physical properties and consequently, evidence for their impact on phenology remains scarce (Badeck et al., 2004). Photoperiod is an important trigger of plant phenology, but regrettably the relative importance of this environmental cue with respect to temperature has been established in only a few wild species (Hunter & Lechowicz, 1992; Kramer, 1994). The balance between rainfall and evaporation plays a key role in ecosystem functioning in many regions of the planet (e.g. in Mediterranean biomes). However, precipitation has received little attention in studies of historical records of plant phenology (Sparks et al., 1997; Pen˜ uelas et al., 2002, 2004; Williams & Abberton,

2004; Gordo & Sanz, 2005; Estrella & Menzel, 2006), despite of precipitation patterns will change in. the future (Solomon et al., 2007) and thus, they could promote alterations in plant phenological patterns as well. Furthermore, precipitation effect in plant phenology is complex and difficult to forecast due to its close relationship with soil moisture. For instance, rainy autumns are related to earlier springs in the following year in some ecosystems (Sparks et al., 1997; Penuelas et al., 2004; but see Cayan et al., 2001). This fact suggests that precipitation may affect individuals even much time later than the last rainy day. Similarly, temperature may reveal its effect with some delay, e.g. through chilling requirements during the winter to break bud dormancy. Such temporal gap between plant phenotype expression (i.e. a certain phenological date) and the potential effect of some of its climatic triggers requires a view beyond the present to assess such potential carry-over effects of climate. Plant phenology responds to the stimuli imposed by local weather, but many studies have also demonstrated a connection to large-scale atmospheric circulation patterns. In Europe, when the North Atlantic Oscillation (NAO) index is positive, spring advances (Post & Stenseth, 1999; Chmielewski & Rötzer, 2001; Post et al., 2001; Scheifinger et al., 2002; Menzel, 2003; Stockli & Vidale, 2004; Menzel et al., 2005b; Ahas & Aasa, 2006; Nordli et al., 2008). This relationship is likely mediated by NAO effect in local weather, e.g. through temperature and rainfall. For instance, positive values of NAO from December to March are related to warm and wet springs in central and northern Europe, but cold and dry springs in the Mediterranean Basin (Ottersen et al., 2001).

Conclusion

Global change experiments have also documented the influence of increasing temperature, shifting precipitation and other aspects of global change, such as rising CO_2 concentrations on the timing of species and ecosystem level phenology. The recent advances in species-specific phenological models now enable researchers to scale from local to regional shifts in phenology as well as to predict shifts in species ranges in response to changing environmental conditions and use historical records of harvest dates to reconstruct past climate Comparing the findings of multiple approaches is necessary to understand how phenology will shift in response to different aspects of global change, and to identify the processes that scale between species and ecosystem phenology. A meta-database of existing networks shows that most observation networks are located in temperate ecosystems and that long-term phenological observations in the tropics are lacking. Accurate phenology predictions would increase the accuracy of the predictions of ecosystems productivity and gas exchanges with the atmosphere, better understanding of population dynamics in multi-species interactions systems. This could help in better management of agriculture and forests.

Selected References

1. Arora, V.K. and Boer, G.J. (2005) A parameterization of leaf phenology for the terrestrial ecosystem component of climate models. Global Change Biol. 11, 39–59
2. Davis CC, Willis CG, Primack RB, Miller-Rushing AJ (2010). The importance of phylogeny to the study of phenological response to global climate change. Philosophical Transactions of the Royal Society B: Biological Sciences, 365, 3201–3213.
3. Estrella, N. and Menzel, A. (2006) Responses of leaf colouring of four deciduous tree species to climate and weather in Germany. Climate Res. 321, 253–267
4. Menzel, A. (2003) Plant phenological anomalies in Germany and their relation to air temperature and NAO. Climatic Change 57, 243–263
5. Menzel, A. et al. (2006) European phenological response to climate change matches the warming pattern. Global Change Biol. 12, 1969–1976
6. Ollerton J, Lack AJ (1992) Flowering phenology: an example of relaxation of natural selection? Trends in Ecology & Evolution, 7, 274–276.
7. Parmesan, C. and Yohe, G. (2003). A globally coherent fingerprint of climate change impacts across natural systems. Nature 421, 37–42
8. Parmesan C (2007) Influences of species, latitudes and methodologies on estimates of phenological response to global warming. Global Change Biology, 13, 1860–1872.
9. Penuelas, J. et al. (2004) Complex spatiotemporal phenological shifts as a response to rainfall changes. New Phytol. 161, 837–846
10. Scheifinger, H. et al. (2003) Trends of spring time frost events and phenological dates in Central Europe. Theor. Appl. Climatol. 74, 41–51
11. Sparks, T.H. and Carey, P.D. (1995). The responses of species to climate over two centuries: an analysis of the Marsham phenological record, 1736-1947. J. Ecol. 83, 321–329
12. Sparks, T.H. and Tryjanowski, P. (2005). The detection of climate impacts: Some methodological considerations. Int. J. Climatol. 25, 271–277
13. Thackeray SJ, Sparks TH, Frederickson M, Burthe S, Bacon PJ, Bell JR et al. (2010) Trophic level asynchrony in rates of phenological change for marine, freshwater and terrestrial environments. Global Change Biology, 16, 3304–3313.
14. Walther, G.R., Post, E.,Convey,P., Menzel, A.,Parmesan, C., Beebee, T. J. C., Fromentin, J. M., Hoegh-Guldberg, O. & Bairlein, F. 2002 Ecological responses to recent climate change. Nature 416, 389–395. (doi:10.1038/416389a.)
15. Yang LH, Rudolf VHW (2010) Phenology, ontogeny and the effects of climate change on the timing of species interactions. Ecology Letters, 13, 1–10.

ISBN-13: 978-1533168078
ISBN-10: 1533168075

Chemistry of Industrial Globalization, Environmental Pollution and its Chem-Biological Significance

February 2016

Industrial Globalization and Environmental Awareness

Proceedings of the NATIONAL SEMINAR held by
Department of Chemistry, Government PG College, Ambala Cantt - Haryana - India

Green Computing: Heading Towards a Better Future

Nisha Saini

Assistant Professor, PGGC-11, Chandigarh
26nisha1992@gmail.com

Abstract

Technology is the branch of knowledge that deals with the creation and use of technical means and their interrelation with life, society, and the environment, drawing upon such subjects as industrial arts, engineering, applied science and pure science. Technology has influenced the life of humans in numerous ways. It has also laid some negative impacts on the environment. Green computing has emerged as one of the primary solutions to reduce the impacts of technology on environment. Green computing aims at providing alternate ways for the manufacturing as well as dumping of technical components so that the harm caused can be minimized. This paper aims at presenting various green computing techniques and methods that can be used to protect the environment.

1. INTRODUCTION

The field of "green technology' encompasses a broad range of subjects — from new energy-generation techniques to the study of advanced materials to be used in our daily life. Green technology focuses on reducing the environmental impact of industrial processes and innovative technologies caused by the Earth's growing population. It has taken upon itself the goal to provide society's needs in ways that do not damage the natural resources. This means creating fully recyclable products, reducing pollution, proposing alternative technologies in various fields, and creating a center of economic activity around technologies that benefit the environment. The main objective of this technology is to study and practice computing resources efficient and eco-friendly [1]. Maximizing the energy efficiency and to promote biodegradability are the primary focus of this technology. Due to pollutants generated by it and the steady increase in rates, energy consumption is causing serious environmental and economic problems. Regarding energy efficiency, a branch of Green IT named energy-aware computing has evolved. Green computing is very much essential for the future world. It is required to make our self and our environment healthy. It can be defined as responsibly utilizing the resources available. Many computers are produced from many hazardous materials like cadmium, mercury and other toxic substances. While disposing the computers, it will lead to pollution and affect the environment to a great extent. This field encompasses a broad range from new generation techniques to the study of advanced materials to be used in daily life. Bringing it to practice will address many problems that are being a threat to human life and our environment [2].

2. HISTORY OF GREEN COMPUTING

The launch of Energy Star program in 1992 by U.S Environmental Protection Agency. Energy Star is a kind of label awarded to computers and other electronics products. Energy Star program minimizing the use of energy while maximizing efficiency. One of the first approaches towards green computing was sleep mode function in computers. Sleep Mode function which places a computer on standby mode to a pre-set period of time. According to Wikipedia "The Swedish organization TCO development launch the TCO certification program to promote a low magnetic and electrical emission from Cathode Ray Tube (CRT) based computer display; this program was later expanded to include criteria on energy consumption, ergonomics and the use of hazardous material in construction" [3].

3. NEED OF GREEN COMPUTING

Nowadays computer is the basic need of every human. A computer made our life easier and saves a lot of time and human efforts, but the use of computer also increase power consumption and also generate a greater amount of heat. Greater power consumption and greater heat generation means greater emission of greenhouse gases like Carbon Dioxide (CO_2) that has various harmful impacts on our environment and natural resources. This is because we are not aware about the harmful impacts of the use of computer on environment [4]. Personal computers and data centers consume a lot of energy which use various old techniques and they don't have sufficient cooling systems. Resultant is the polluted environment. There are various reasons for the use of green computing are [5]:

A. Computers and electronic devices consume a lot of electricity that have some harmful impact on our environment. It produces air pollution, Land pollution and water pollution. Electricity generated through Fossil Fuel power plants release air pollution and requires a lot of water that effect our environment like climate change, acid rain (pH<5), ozone(O_3) and air toxic.

B. Most of electronic devices generate a lot of heat which cause the emission of CO_2. CO_2 is one of the greenhouse gases, warming the earth surface to higher temperature by reducing outward radiation. With the rapidly increasing of carbon Dioxide the rate of global warming became increase causing and through anthropogenic climate change.

C. While disposing of computers and it resources produces a lot of hazardous waste that really damage our environment. It also releases heavy metal like lead (Pb), mercury (Hg), cadmium (Cd) into air.

D. The manufacturing of computers product release heavily on the use of toxic comical for electrical insulation, soldering, and fire protection. Expose the comical fumes over the long term can cause cancer, cause miscarriages.

All these causes can be reduced using one concept i.e. "Green computing". Now we have needed to implement the green computing on various electronic and electrical devices to save our environment from these harmful impacts.

4. TECHNIQUES OF GREEN COMPUTING

Various green computing techniques can be stated as follows [6]:

- Carbon-free computing

One of the ideas is to reduce the carbon footprint of users - the amount of greenhouse gases produced, measured in units of carbon dioxide (CO_2). Greenhouse gases naturally blanket the Earth and are responsible for its more or less stable temperature. An increase in the concentration of the main greenhouse gases such as carbon dioxide, methane, nitrous oxide, and fluorocarbons is believed to be responsible for Earth's increasing temperature, which could lead to severe floods and droughts, rising sea levels, and other environmental effects, affecting both life and the world's economy

- Solar Computing

Solar computing can be seen as one of the significant initiatives for green-computing projects. Solar cells fit power-efficient silicon, platform, and system technologies and enable the company to develop fully solar-powered devices that are nonpolluting, silent, and highly reliable. Solar cells require very little maintenance throughout their lifetime, and once initial installation costs are covered, they provide energy at virtually no cost. Worldwide production of solar cells has increased rapidly over the last few years; and as more governments begin to recognize the benefits of solar power, and the development of photovoltaic technologies goes on, costs are expected to continue to decline.

- Lead-Free and RoHS computing

In February 2003, the European Union adopted the Restriction of Hazardous Substances Directive (RoHS). The legislation restricts the use of six hazardous materials in the manufacture of various types of electronic and electrical equipment. The directive is closely linked with the Waste Electrical and Electronic Equipment Directive (WEEE), which sets collection, recycling, and recovery targets for electrical goods and is part of a legislative initiative that aims to reduce the huge amounts of toxic e-waste.

- Vision through the pc-1

The VIA pc-1 initiative seeks to enable the next 1 billion people to get connected, by providing wider access to computing and communications technologies. The company is concentrating on empowering new, emerging markets, looking at models that reach beyond individual ownership of a PC, such as local pay-for-use facilities. Helping to build skills and literacy throughout the world and incorporating and preserving cultural content are goals now within our grasp. Information is the oxygen to nurturing social mobility, economic equality and development, and global democracy. Providing not just the tools and the know-how, but the support and the maintenance, is all part of what makes pc-1 the next generation of information technology, the next generation of global development" [7]. Among the company's projects under the pc-1 program are the —tuXlab‖ computer center in South Africa and an ICT Training Center in Vietnam.

5. CONCLUSION

With the advancement of technology, the world has become a better place to live because of the facilities given by it in various fields. Along with these facilities, the negative impacts of technology can't be overlooked. The presented paper aims at reducing the negative impacts of technology and thus trying to make the environment free from technological hazards. The paper has put a spotlight on various techniques of green computing that can be implemented in future.

6. REFERENCES

1) S.V.S.S. Lakshmi, I Sri Lalita Sarwani, Nalini Tuveera, "A Study on Green Computing: The Future Computing and Eco-Friendly Technology". International Journal of Engineering Research and Applications ISSN: 2248-9622, Vol.2, Issue 4, July-August 2012, Pp.1282-1285.

2) Shalabh Aggarwal, Arnab Data, Asoke Nath,"Impact of Green Computing in IT industry To Make Eco-Friendly Environment", Journal of Global Research in Computer Science, ISSN: 2229-371X, Vol.5, No. 4, April 2014, Pp. 5-10.

3) A. Mala, C. Uma Rani, L. Ganesan, "Green Computing: Issues on the Monitor of Personal Computers", International Journal of Engineering And Sciences, ISSN(e):2278-4721,ISSN(p):2319-6483,vol.3,issue 2,May 2013,Pp 31-36.

4) Dr. Pardeep Mittal, Navdeep Kaur, "Green Computing Need and Implementation", International Journal of Advanced Research in Computer Engineering and Technology, ISSN: 2278-1323, Vol.2, ISSUE 3, March 2013, Pp. 1200-1203.

5) Masood Anwar, Syed Furqan Qadari, Ahsan Raza Sattar, "Green computing and Energy Consumption Issue in the Modern Age", Journal of Computer Engineering, ISSN:2278-9359, December 2012, Pp. 14-18.

6) Gaurav Jindal, Manisha Gupta,"Green Computing: Future of Computers", International Journal of Engineering Research in Management and Technology, ISSN: 2278-9359, December 2012, Pp. 14-18.

7) Sharmila Shinde, Simantini Nalwade, Ajay Nalwade,"Green Computing: Go Green and Save Energy", International Journal of Advanced Research in Computers Sciences and Software Engineering, ISSN: 2277 128X, Vol.3, Issue 7, July 2013, Pp. 1033-1037.

ISBN-13: 978-1533168078
ISBN-10: 1533168075

Chemistry of Industrial Globalization, Environmental Pollution and its Chem-Biological Significance

February 2016

Industrial Globalization and Environmental Awareness

Proceedings of the NATIONAL SEMINAR held by
Department of Chemistry, Government PG College, Ambala Cantt - Haryana - India

SYNTHESIS & CATALYSIS OF POLYSTYRENE-ANCHORED U(VI) & Mo(VI) COMPLEXES

Jai Pal

Department of Chemistry, S.D. College-Ambala Cantt(Haryana) India
Email: drsaharan19@gmail.com

Abstract

[PS-LUO$_2$.DMF & PS-LMoO$_2$DMF] were synthesized by the reaction of tridentate dibasic PS-LH$_2$ and metal salts. Catalysts were characterized by elemental analyses, IR, GC-MS. The oxidation of various alcohols such as benzyl alcohol, 2-butanol, and 2-propanol was investigated using the supported metal complexes in presence of molecular oxygen as the oxidant. The swelling studies were done by using different solvents. Kinetic data indicate that the catalysts could be recycled without significant degradation of polymer matrix. Reaction temperature and concentration of active metal center in the catalyst have great influence on the reaction rate. catalyst could be recycled at least 5 times without loss of activity.

Keywords: Synthesis, Supported Metal Complexes, Oxidation of Alcohols, Catalysis

Introduction:

The field of polymer metal complexes as catalysts in organic synthesis has remained active ever, since the first example of catalysis by a polymer linked transition metal compound were reported in 1952 by Sherrington and others have reviewed in detail the various applications of supported catalysts in organic transformations [1-3]. It is now well recognized that the advantages derived from the use of functionalized macromolecules are generally associated with a simplification of the work-up, easy separation of products from the reaction mixture, recovery and recycling of the catalyst, and the possibility of these catalysts being used in continuous flow systems or in automated synthesis [4-5]. Polymer supported metal complexes are extensively used as oxidizing agents, reducing agents, photosensitizes, agriculturally and pharmacologically active reagents [6-8]. The applications of polymer-metal complexes in the field of catalysis have been widely investigated. Polymer-metal complexes are marked by their use as immobilized reagents which are useful for industrial purposes. Among organic polymer Chloromethylated polystyrene cross linked with divinyl benzene has been the polymer of choice with a wide range of functional groups incorporated in it to bind the metal into the polymer. The basic polymer backbone being chemically inert the polar properties can be modified by controlled functionalization. Polystyrene can be functionalized easily, because it incorporates aryl groups. In polystyrene based system the ability to control the pore size, either through the amount of cross-linking agent or by the choice of a solvent allows some steric selectivity which is not possible in homogeneous system. Oxidation with molecular oxygen catalyzed by transition metal complexes provides an attractive route for the preparation of synthetic intermediates and other oxygen containing organic substrates without the use of environmentally hazardous oxidants [9].

Acknowledgments: author is thankful to UGC-New Delhi, India for providing Minor Research Project Grant. The author is also thankful to Principal SD College-Ambala Cantt(Hr) for providing research facilities.

Refrences

1. Dalal, M.K., Upadhyay, M.J. and Ram, R.N. Catalytic Activity of Polymer Bound Ru (III)-EDTA Complex. *Journal of Molecular Catalysis A: Chemical*, **142**(1999)325-332.
2. Chettiar, K.S. and Sreekumar, K. Polymer Supported Thiosemicarbazone Transition Metal Complex. *Polymer International,* **48**(1999) 455-450.
3. Manyar, H.G., Chaure, G.S. and Kumar, A. Supported Polyperoxometallates, Highly Selective Catalyst in Oxidation of Alcohols to Aldehydes. *Journal of Molecular Catalysis A: Chemical,* **243**(2006) 244-252.
4. Zhang, Z. and Wang, Z. Studies on Pd/NiFe$_2$O$_4$ Catalyzed Ligand-Free Suzuki Reaction in Aqueous Phase: Synthesis of Biaryls, Terphenyls and Polyaryls. The Journal of Organic Chemistry, **71**(2006) 7485-7487.
5. Dalal, M.K. and Ram, R.N. Catalytic Activity of Polymer-Bound Ru(III)-EDTA Complex. *Bulletin of Materials Science*, **24** (2001) 237-241.

6. Hartley, F.R., Murray, S.G. and Nicholson, P.N. γ-Radiation Produced Supported Metal Complexes a New Generation of a Catalyst. *Journal of Organometallic Chemistry,* **231**(1982) 369-372.

7. Hartley, F.R., Murray, S.G. and Nicholson, P.N. Wacker Chemistry with Solid Catalysts. *Organometallic Chemistry*, **16**(1982), 363.

8. Chatterjee, D., Bajai, H.C., Das, A. and Shatt, K. Studies of Some New Schiff Base Complexes of Ruthenium. *Journal of Molecular Catalysis*, **92**(1994) L235-L238.

9. Sherrington, D.C. Polymer Supported Metal Complex Alkene Epoxidation Catalysts. *Catalysis Today*, **57**(2000) 87- 104.

ISBN-13: 978-1533168078
ISBN-10: 1533168075

Chemistry of Industrial Globalization, Environmental Pollution and its Chem-Biological Significance February 2016

Industrial Globalization and Environmental Awareness

Proceedings of the NATIONAL SEMINAR held by
Department of Chemistry, Government PG College, Ambala Cantt - Haryana - India

"Effects of Toxic Chemicals on Environmental and Human Health"

Dr AVTAR SINGH RAHI

Head and Associate Professor, Department of Chemistry,
Government Post-Graduate College, Ambala Cantt. (India)
rahiavtaar@gmail.com

Chemicals play a major role in our lives and environment. Research in chemistry has brought a dramatic increase in the invention, production and consumption of enormous chemicals. Although some of these everyday chemicals like pharmaceutical and cosmetic products, food additives, etc. might not be of direct environmental concern but numerous other chemicals are continuously introduced into the environment in large quantities like agro-chemicals, detergents, dyes, perfumes, deodorants, textiles, plastics, paints and varnishes, construction byproducts, solvents, etc. Researchers do not focus on possible environmental and health effects. Industrial enterprises also do not care for these effects but try to reap the maximum financial benefits. Now the research and care has been initiated to discover the impact and effects of these toxic and hazardous chemicals on environment and human-plants-animals health. It is not only the larger quantities but smaller and even the minute quantities can make adverse changes. Adverse chemicals impact can reach to any Human population, geographic locations, employment, diet, etc. Protecting against potential health hazards requires knowledge about commercial chemicals, various mixtures, byproducts, etc. not only to common man but to researchers, scientists, administrators and policy makers also. Not the use but efficient use of chemicals can solve the problem of environmental contamination and pollution. One should know more about these chemicals' persistence and fate in the environment. What effects they will have and most important, how to reduce the risk these chemicals may pose. Research, production of chemicals and byproducts and consumption should be the part of monitoring. Every step should accompany studies on environmental and health impact. Chemicals we invent, effects less adverse but byproducts or side-products which are produced without our wish or by-chance, have adverse effects. That invention or production termed better which produces anti-dotes first and then chemicals. Environment is to enjoy and not to contaminate or pollute because every action has reaction. What we have injected or transmitted to environment, we are getting back and more disregards to nature may lead to very dangerous results, even extinctions.

ISBN-13: 978-1533168078
ISBN-10: 1533168075

Chemistry of Industrial Globalization, Environmental Pollution and its Chem-Biological Significance

February 2016

Industrial Globalization and Environmental Awareness

Proceedings of the NATIONAL SEMINAR held by

Department of Chemistry, Government PG College, Ambala Cantt - Haryana - India

Water Pollution from Pulp and Paper Mills

Dr. Ashima

Department of Chemistry, ASSM College Mukandpur. Punjab.
E-mail: passiashima@gmail.com

ABSTRACT

Pulp and paper mills are considered one of the most polluting industries worldwide. Paper making process demands large amount of fresh water and produces enormous quantities of wastewater. The wastewater is contaminated by a number of organic and inorganic chemicals including lignin, cellulosic compounds, phenols, mercaptans, sulfides and chlorinated compounds. The amount and characteristics of wastewater depends upon scale of operation, raw materials used and the process employed. Biochemical oxygen demand (BOD) and chemical oxygen demand (COD) of the waste stream may be in the ranges of 10 – 40 kg/t and 20 – 200 kg/t of air dried pulp, respectively. Generally, small pulp and paper mills (< 100 t/d) generate smaller quantities of wastewater but with higher pollution load. In a pulp and paper mill, several unit processes are used for manufacturing the final product. Pulping and bleaching processes are the two major sources of highly polluted water. The wastewater from pulp and paper mills has generally low biodegradability due to the presence of recalcitrant compounds. These pollutants are resistant to conventional biological treatment processes and inhibit the purification of the wastewater. Disposal of such wastewater in aquatic bodies can have severe adverse impacts on the living organisms. This paper looks at the sources of pollution from a paper and pulp industry and steps being implemented to tackle this challenge. The environmental impacts of the wastewater discharge in natural water bodies are discussed. To mitigate the pollution caused by pulp and paper mill wastewater, the possible ways of wastewater treatment and recycling of treated water are also suggested.

Keywords: pulp and paper mills, water pollution, waste water paper mills,

ISBN-13: 978-1533168078
ISBN-10: 1533168075

Chemistry of Industrial Globalization, Environmental Pollution and its Chem-Biological Significance February 2016

Industrial Globalization and Environmental Awareness

Proceedings of the NATIONAL SEMINAR held by

Department of Chemistry, Government PG College, Ambala Cantt - Haryana - India

Cause and Effects of Global Warming

Harish Soni[1] and Deepak Kumar[2]

[1]Department of Chemistry, Government College(PG), Ambala Cantt, Haryana, India
Email- harishkumar484@yahoo.com
[2]Department of Mathematics, G.M.N (PG) College , Ambala Cantt, Haryana, India
Email- sonideepak2210@gmail.com

Abstract

Global warming is the product of green house effect. Global warming is the term used to describe a gradual increase in the average temperature of the earth's atmosphere and its oceans, a change that is believed to be permanently changing the earth climate. It is due the increased carbon dioxide and green house gases in air due to pollution caused by human and it is predicted that it will cause variety of negative effects. The cause of global warming is the increasing quantity of green house gases in the our atmosphere produced by human activities, like burning of fossil fuels or deforestation. Finally, this Paper presents the Cause and Effects of Global warming.

Keywords: Global Warming, Cause, Effect, Atmosphere, Gases,

ISBN-13: 978-1533168078
ISBN-10: 1533168075

Chemistry of Industrial Globalization, Environmental Pollution and its Chem-Biological Significance February 2016

Industrial Globalization and Environmental Awareness

Proceedings of the NATIONAL SEMINAR held by

Department of Chemistry, Government PG College, Ambala Cantt - Haryana - India

ROLE OF CHEMISTRY IN ENVIRONMENTAL PROTECTION

POOJA SHARMA

Department of Chemistry, S D College-Ambala Cantt(Haryana)133001
Email id: poojashrm603@gmail.com

Chemistry is one of the oldest branches of science; the human beings had ever come across. It has consistently contributed towards meeting the human needs from the dawn of civilization. However, its role has multiplied since the inception of industrial revolution. Although anthropogenic activities have made the human life comfortable and even luxurious yet their impacts on the physical, biological and socio-economic environments had been destructive. Numerous kinds of chemicals have engulfed us and our environment. Modern chemistry has leading role in sculpting the present as well as future of human lifestyle. It is serving the man and other biodiversity by providing countless products in every sphere of life. At the same time it is playing villain role in the destruction of environment at an alarming rate. Today the world is confronted with heinous environmental issues hitherto unknown to the living beings mostly triggered by chemicals. Thousands of chemicals are used in industrial products, agricultural chemicals, persistent organic pollutants, freezers, pharmaceuticals, chemical and radiological warfare, construction industry, synthetic materials, electrical goods, medical gadgets etc. Some natural sources of chemicals are acid rains, volcanic eruptions, eutrophication and photochemical smog. The fact of matter is that chemicals are being consistently added into atmosphere, biosphere and lithosphere. For the sustainable environment it is imperative that the chemicals must not be added into human environment beyond its carrying capacity. It is responsibility of chemists to introduce environmentally benign and biodegradable chemicals. All types of chemistry need to be green and environment friendly. The scientists and engineers should develop chemicals and technologies which do not harm the living creatures during any stage of their life-cycle

Keywords: Modern Chemistry, Biodegradable chemicals, Eutrophication

ISBN-13: 978-1533168078
ISBN-10: 1533168075

Chemistry of Industrial Globalization, Environmental Pollution and its Chem-Biological Significance

February 2016

Industrial Globalization and Environmental Awareness

Proceedings of the NATIONAL SEMINAR held by

Department of Chemistry, Government PG College, Ambala Cantt - Haryana - India

Current Scenario of E-waste Management in India

Dr Alka

Assistant Professor in Physics, Govt. College, Barwala (Panchkula)

E-mail: alkagc1@yahoo.co.in

Abstract

The electronic waste also known as e-waste consists of broken or discarded electrical or electronic equipment. The widespread use of electronic goods and dumping of electronic goods by the developed countries has brought the e-waste problem in India to an acute crisis. The toxic and hazardous substances present in e-waste pose a danger to human health as well as environment. Concerted efforts by various players in the electronics industry, academic community and the government are required to implement a systematic e=waste management system in India. This paper attempts to provide a brief insight into the concept of e-waste in India, the environmental and health hazards attached to it, the methods being used for e-waste management and the legislation work done regarding e-waste in India.

ISBN-13: 978-1533168078
ISBN-10: 1533168075

Chemistry of Industrial Globalization, Environmental Pollution and its Chem-Biological Significance

February 2016

Industrial Globalization and Environmental Awareness

Proceedings of the NATIONAL SEMINAR held by

Department of Chemistry, Government PG College, Ambala Cantt - Haryana - India

Volumetric and Ultrasonic Studies of Binary Liquid Mixtures of Alkoxypropanols with Cyclic Amide at Different Temperatures

Dr. Anil Kumar

HOD Chemistry, Arya P.G. College, Panipat

E- mail: kumar_anilab@rediffmail.com

Abstract

Densities and speeds of sound for the binary liquid mixtures of Alkoxypropanols with cyclic amide have been measured using an Anton Paar DSA 5000 instrument at temperature from 288.15 K to 308.15 K. The excess molar volumes V_M^E, and excess molar isentropic compressibility $K_S{}^E{}_m$ were calculated from Experimental data . The computed quantities were fitted to Redilich – Kister equation to derive the coefficients and estimate the standard error values. The ultrasonic speed value have been combined with those of the excess molar volumes converted to densities to give estimates of the product Ks,m of the molar volume and the isentropic compressibility, Ks and the excess quantity $K_S{}^E{}_m$ The sign and magnitude of V_M^E and $K_S{}^E{}_m$ were used to analyze the behaviour of the components in terms of intermolecular interactions.

ISBN-13: 978-1533168078
ISBN-10: 1533168075

Chemistry of Industrial Globalization, Environmental Pollution and its Chem-Biological Significance

February 2016

Industrial Globalization and Environmental Awareness

Proceedings of the NATIONAL SEMINAR held by
Department of Chemistry, Government PG College, Ambala Cantt - Haryana - India

Synthesis of Some Novel N-aryl- 2-mecrcaptoimidazoles as Potential Antimicrobial Agents

Vinod Kumar[*a] , **Kuldeep Singh**[a,b] , **Devinder Kumar**[b], **Mayank Kinger**[a]

[a]Department of Chemistry, M. M. University, Mullana-Ambala, 133207, India
[b]Department of Chemistry, Guru Jambheshwar University of Science and Technology, Hisar
* Corresponding author E-mail: vinodbatan@gmail.com

Abstract

The major cause of morbidity and mortality world-wide is due to increasing resistance of microorganisms to currently available antimicrobial drugs. In order to explore antimicrobial potential, some novel N-aryl- 2-mecrcaptoimidazoles have been synthesized. To attain target, firstly, the reaction of 3-aminoacetophenone was performed with different phenacyl bromides which results the corresponding anilino compounds which on further treatment with potassium thiocyanate gave 1-[3- (2-mercapto- 4-phenyl- 1H-imidazol- 1-yl)phenyl]ethanones exclusively. All the synthesized compounds were characterized on the basis of their 1 H, 13 C-NMR and HRMS spectral data. Antimicrobial potential of the synthesized compounds was evaluated from their in vitro antimicrobial screening against Gram-positive, Gram-negative bacterial and fungal stains using the standard drugs, ciprofloxacin and fluconazole.

Keywords: Mercaptoimidazoles, antibacterial, antifungal, NMR spectroscopy.

ISBN-13: 978-1533168078
ISBN-10: 1533168075

Chemistry of Industrial Globalization, Environmental Pollution and its Chem-Biological Significance

February 2016

Industrial Globalization and Environmental Awareness

Proceedings of the NATIONAL SEMINAR held by

Department of Chemistry, Government PG College, Ambala Cantt - Haryana - India

Physico-Chemical Studies of Nonlinear Optical Guest-Host Polymeric Thin Films

Sushil Kumar[1] & Sanjiv Arora[2]

[1]Department of Chemistry, S. D. College, Ambala Cantt-133 001, India
[2]Department of Chemistry, Kurukshetra University, Kurukshetra - 136 119, India
E-mail- drsushilgoswami@yahoo.com

We discuss here the physico-chemical studies of prepared guest-host systems obtained by physical mixing of different concentrations (2, 4, 6, 8, 10 % by weight) of azobenzene derivative (nonlinear optical chromophore) as guest in polymethyl methacrylate (PMMA) as host. First, we have characterized various guest-host systems by IR spectra, DSC and TGA techniques. After characterization, the vacuum-deposited thin films of these guest-host blends were used to study their optical properties. The efforts were made to study the effect of size and concentration of the dopant chromophore on the orientational stability of the guest-host systems. The results suggested that aligned dipoles in these polymeric thin films possess suitable temporal stability.

ISBN-13: 978-1533168078
ISBN-10: 1533168075

Chemistry of Industrial Globalization, Environmental Pollution and its Chem-Biological Significance
February 2016

Industrial Globalization and Environmental Awareness
Proceedings of the NATIONAL SEMINAR held by
Department of Chemistry, Government PG College, Ambala Cantt - Haryana - India

Recycling and the Environment

Reena, Assistant Professor in Physics, Govt. College for Women, Tosham
Anil Kumar, Asstt. Prof. of Physics, Govt. College for Women, Bawani Khera

Abstract

Recycling is the process to convert waste materials into reusable material to prevent waste of potentially useful materials, reduce air and water pollution by reducing the need for conventional waste disposal and lower green house gas emissions as compared to plastic production. The ways by which recycling helps environment could be reducing landfill, reducing energy consumption, decrease pollution. In this paper we discussed that biodegradable are slow to decompose so avoid them. Grow more trees in order to absorb the carbon. Main benefit of recycling is that there is significant cost savings .Recycling leaves and grass are great to make compost. At last recycling is just one of many ways that we can help the Environment. This will help produce a better environment for many generations.

Role of needs to explore different applications of Technology

Anil Kumar, Asstt. Prof. of Physics, Govt. College for Women Bawani Khera
Reena, Asstt. Prof. of Physics, Govt. College for Women, Tosham

Abstract

Role of Technology is increasing day by day, we all depend on technology and we use various technologies to accomplish specific tasks in our lives. Today we have various ensuing technologies which impact our lives in different ways. Technology is being implemented in almost every section of our lives. Technology is playing immense role in education, health, business, communication, agriculture and many more. So embracing it and learning how to use technology in whatever we do is very important and recommended. As the world keeps on developing, technology will be changed, what is working today might not work and not be efficient tomorrow. So it is better to stay up-to- date with new emerging technologies and learn how to embrace and use them in your daily life.

ISBN-13: 978-1533168078
ISBN-10: 1533168075

Chemistry of Industrial Globalization, Environmental Pollution and its Chem-Biological Significance

February 2016

Industrial Globalization and Environmental Awareness

Proceedings of the NATIONAL SEMINAR held by
Department of Chemistry, Government PG College, Ambala Cantt - Haryana - India

Global Warming and Its Solutions

Parisha

Assistant Professor in Computer Science,
Government College, Bhiwani.

This Abstract discusses about the Global Warming and its solutions. Global Warming is the increase of Earth's average surface temperature due to effect of Greenhouse gases, such as carbon dioxide emissions from burning fossil fuels or from deforestation, which trap heat that would otherwise escape from Earth. In this paper, we discussed solutions to Global warming such as set limits on global warming pollution, drive smarter cars, and create green homes and buildings to save energy, Invest in green jobs and clean energy. These steps can be followed to reduce the energy use, improve efficiency and help end global warming. At last we can conclude for reducing Global Warming, we should buy energy efficient appliances and replace our lights bulbs with compact fluorescent bulbs. CFLs lower your energy bills and keep carbon dioxide out of the air. Use public transport and walking which will help in reducing pollution in the air.

ISBN-13: 978-1533168078
ISBN-10: 1533168075

Chemistry of Industrial Globalization, Environmental Pollution and its Chem-Biological Significance

February 2016

Industrial Globalization and Environmental Awareness

Proceedings of the NATIONAL SEMINAR held by

Department of Chemistry, Government PG College, Ambala Cantt - Haryana - India

Clutter Processing for Highly Sensitive Radar Using Wavelet Transforms

Praveen Kumar Yadav

Assistant Professor in Physics,
Government College, Kosli (Rewari)

Abstract

A wavelet transform based clutter processing approach for signal processing increase the accuracy for detection of the target. The addition of digital processing to radars, nowadays increased the radar's sensitivity. This requires more robust clutter processing to maintain optimal system performance. In 1982, Jean Morlet a French geophysicist, introduced the concept of a `wavelet transforms'. The wavelet means small wave and the study of wavelet transform is a new tool for signal analysis. Also, Alex Grossmann theoretical physicists studied inverse formula for the wavelet transform. The joint collaboration of Morlet and Grossmann yielded a detailed mathematical study of the continuous wavelet transforms and their various applications. Similar results had already been obtained in 1950's by Calderon, Littlewood, Paley and Franklin. However, the rediscovery of the old concepts provided a new method for decomposing a function or a signal. Wavelet analysis is originally introduced in order to improve seismic signal analysis by switching from shortime Fourier analysis to new better algorithms to detect and analyze abrupt changes in signals. In time-frequency analysis of a signal, the classical Fourier transform analysis is inadequate because Fourier transform of a signal does not contain any local information. This is the major drawback of the Fourier transform. To overcome this drawback, Dennis Gabor in 1946, first introduced the short-time Fourier transform known later as Gabor transform. The modern applications of wavelet theory are diverse as wave propagation, data compression, signal processing, image processing, pattern recognition, computer graphics, the detection of aircraft and submarines and some other medical image technology etc.

ISBN-13: 978-1533168078
ISBN-10: 1533168075

Chemistry of Industrial Globalization, Environmental Pollution and its Chem-Biological Significance

February 2016

Industrial Globalization and Environmental Awareness

Proceedings of the NATIONAL SEMINAR held by

Department of Chemistry, Government PG College, Ambala Cantt - Haryana - India

Biodiversity Gain with Declining Population Sizes

Dr Anil Jindal[1] and Dr Surender Singh[2]

[1]Asst. Professor of Zoology, R.K.S.D. College, Kaithal
[2]Asst. Professor of Zoology, Govt. P.G. College, Jind

Biodiversity is the variety of life on Earth, it includes all organisms, species, and populations; the genetic variation among the; and their complex assemblages of communities and ecosystems. As new portions of terrestrial wilderness continue to be utilized or modified by human activity, wildlife has less territory, individual species are crowded into smaller spaces, and many of them lose population size until their existence becomes precarious. Many authorities believe that the world's foremost conservation problem is biodiversity loss caused by the extinctions of thousands of species per year. Estimates of huge losses are based on indirect evidence such as the amount of habitat destroyed, pollution, or overexploitation. But, we now have documented records of species extinctions that provide information about diversity loss. By using extinction records for well-known animal groups plus surrogate data, Is how there is no evidence for an unusually high rate of extinction, a mass extinction is not yet underway, and there are indications of a continued biodiversity gain. On the other hand, there is ample evidence to demonstrate the persistence of numerous small populations that are the remnants of once widespread and productive species. These populations represent an extinction debt that will be paid unless they are rescued through present day conservation activity. They constitute the world's true biodiversity problem.

ISBN-13: 978-1533168078
ISBN-10: 1533168075

Chemistry of Industrial Globalization, Environmental Pollution and its Chem-Biological Significance

February 2016

Industrial Globalization and Environmental Awareness

Proceedings of the NATIONAL SEMINAR held by
Department of Chemistry, Government PG College, Ambala Cantt - Haryana - India

Quark Diagram Analysis of Bottom Meson Decays Emitting Two Pseudoscalar Mesons

Maninder Kaur

Department of Physics, Punjabi University, Patiala – 147002
E-mail: maninderphy@gmail.com

In this paper, we investigate phenomenologically two-body weak decays of the bottom mesons emitting two pseudoscalar mesons. Due to the strong interaction interference, like FSI and nonfactorizable contributions, on these processes, it is not possible to calculate their contributions reliably. For instance, weak annihilation and W-exchange contributions, which are naively expected to be suppressed in comparison to the W-emission terms, may become significant due to possible nonfactorizable effects arising through soft-gluon exchange around the weak vertex. Since such effects are not calculable from the first principles, we employ the model independent Quark diagram approach, naively called Quark Diagram Scheme at SU(3) level for various weak quark level processes responsible for these decays.

Decay amplitudes can be expressed independently in terms of possible quark flavour diagrams- like: a) the external W-emission diagram, b) the internal W-emission diagram, c) the W-exchange diagram, d) the W-annihilation, and e) the W-loop diagram, and parameterize their contributions to B meson decays. We construct the weak Hamiltonian responsible for the B decays. Choosing appropriate components of the weak Hamiltonian for these quark level processes, we then obtain several relations among their decay amplitudes in CKM enhanced as well as suppressed modes. We derive corresponding relations among their branching fractions in the QDS using SU(2)-isospin, SU(2)-U spin, flavor SU(3) frameworks to compare with available experimental information.

ISBN-13: 978-1533168078
ISBN-10: 1533168075

Chemistry of Industrial Globalization, Environmental Pollution and its Chem-Biological Significance

February 2016

Industrial Globalization and Environmental Awareness
Proceedings of the NATIONAL SEMINAR held by
Department of Chemistry, Government PG College, Ambala Cantt - Haryana - India

CORPORATE SUSTAINABILITY AUDITING – A LOOK INTO INDIAN CORPORATE SECTOR

Dr. Neeraj Goyal[1] and Dr. Varun Jain[2]
[1]Head Department of Business Management, M M Modi College, Patiala.
[2]Head Department of Mathematics, M M Modi College, Patiala.

Abstract

We are familiar with the term 'Audit' in the sense of examination of financial accounts and records of any organization. Environmental Audit on the other hand refers to a management tool comprising a systematic, documented, periodic and objective evaluation of the performance of the organisation, management system and processes designed to protect the environment with the aim of facilitating management control of practices which may have impact on the environment, and assessing compliance with company policies. It is a programme undertaken to assure compliance with regulatory requirements and corporate guidelines along with application of the best management practices to reduce environmental liabilities and risks, as well as to assure the minimization potential liabilities and risks. The objective of such audit is to measure the impact of an organisation's operations on the wider environment against some predetermined set of criteria and so far as is possible, to account for them. The concept of environmental auditing in India, appears to have first got into meaningful discussions in the beginning of the nineties. This process finally resulted in the issuing of a gazette notification on 13th March 1992 through which submission of the environmental audit reports has been made mandatory for many industries such as Cement factories (above 200 tonnes per day production capacity), Thermal power plants, Fermentation/Distillery factories, Sugar factories, Fertiliser, Pulp & paper Industries (above 30 tonnes per day production), Oil refineries etc. The industries are now required to submit their audit reports to the concerned state pollution control boards on or before 15th day of May every year beginning 1993. It is need of the hour that increased awareness for environment should be translated into reality by practicing environmental audit by Indian corporate sector and thereby making our country clean and green. Present paper is an attempt to look into environmental audit requirements in our country and compliance of these by Indian corporate.

Keywords : Environmental Audit, Environmental Risk Assessment, Environmental Liability, Corporate Audit Regulations, Sustainability Audit.

ISBN-13: 978-1533168078
ISBN-10: 1533168075

Chemistry of Industrial Globalization, Environmental Pollution and its Chem-Biological Significance

February 2016

Industrial Globalization and Environmental Awareness

Proceedings of the NATIONAL SEMINAR held by
Department of Chemistry, Government PG College, Ambala Cantt - Haryana - India

Synthesis of 7-amino- 2,5-diarylpyrazolo[1,5-a]pyrimidines for Antimicrobial Evaluation

Gulshan Singh[a*] , Ranjana Aggarwal[b]

[a]S. D. College (Lahore), Ambala Cantt-133001, Haryana, India
[b]Department of Chemistry, Kurukshetra University, Kurukshetra-136119, Haryana
* E-mail: gulshanbanisingh@yahoo.co.in

Pyrazolo[1,5-a]pyrimidines being analogous to purine bases are found to possess wide applications in the field of medicine and agriculture and can behave as antimetabolite in purine mediated biochemical reactions. Compounds of this class are found as potent cyclooxygenase-2 (COX-2) and cyclin-dependent kinase-2 (CDK2). Antimicrobial activity has also been found to be associated amino substituted pyrazolopyrimidines. Pyrazolo[1,5-a]pyrimidines can be synthesized by condensation of 5-aminopyrazoles with 1,3-diketones. Whereas, treatment of 3-oxo- 3-arylpropanenitrile with 5-aminopyrazoles can avail amino substituted pyrazolo[1,5-a]pyrimidines. 3-oxo- 3-phenylpropanenitrile, 5-amino- 3-phenylpyrazole and PTSA on refluxing for 5 hr in toluene-ethanol afforded 7-amino- 2,5-diphenylpyrazolo[1,5-a]pyrimidine. The structure of the products was established on the basis of vigorous analysis of IR, NMR and mass spectrometry. All synthesized compounds will be screened for antimicrobial activity.

ISBN-13: 978-1533168078
ISBN-10: 1533168075

Chemistry of Industrial Globalization, Environmental Pollution and its Chem-Biological Significance

February 2016

Industrial Globalization and Environmental Awareness

Proceedings of the NATIONAL SEMINAR held by
Department of Chemistry, Government PG College, Ambala Cantt - Haryana - India

Recycling: Use and Impacts

SHASHI SHARMA and LATIKA
Asstt. Prof. G.N.K. College Karnal
email:shashisharma_scientist2007@yahoo.com

Abstract

Recycling is a viable alternative in getting back the usable material and energy. As petroleum prices increases it is becoming more financially problem and valuable sight for recycling polymers rather than produce them from raw materials. In addition to the energy recycling of plastic production and manufacturing products also require energy during use and disposal. A major portion of plastic produced each year is used to make disposable items of packaging or other short-lived products that are discarded within a year of manufacture. To date, the impacts on various collection methods—source-separated curbside, commingled curbside, deposit/return—on the quality of materials destined for recycling have not been formally researched and documented. In fact, rarely is —material quality‖ or the —end-destination‖ of the material considered by government decision-makers when choosing an appropriate recycling system. Impacts on glass, including color mixing, suitability for certain end-uses, and increased operating costs; and Impacts on plastic quality and costs are highlighted in this paper.

ISBN-13: 978-1533168078
ISBN-10: 1533168075

Chemistry of Industrial Globalization, Environmental Pollution and its Chem-Biological Significance February 2016

Industrial Globalization and Environmental Awareness
Proceedings of the NATIONAL SEMINAR held by
Department of Chemistry, Government PG College, Ambala Cantt - Haryana - India

Global Warming: Causes and Effects

Dr Rajendra Swain[1] and Narinder Anchal[2]

[1]Asstt. Professor, Dept. of Chemistry, PG Govt. College for Girls, Sector -42, Chandigarh
[2]Assoc. Professor, Dept. of Chemistry, PG Govt. College for Girls, Sector -42, Chandigarh
Email: drrajendra252@gmail.com **Email**: nanchal63@yahoo.com

Global warming is damaging the Earth's climate as well as the physical environment. One of the most visible effects of global warming is harming the environment in several ways including Desertification, Increased melting of snow and ice, Sea level rise and Stronger hurricanes and cyclones. Substantial scientific evidence indicates that an increase in the global average temperature of more than 2°F above where we are today poses severe risks to natural systems and human health and well-being. To avoid this level of warming, steps taken to reduce heat-trapping emissions by 2050. Delay in taking such action will require much sharper cuts later, which would likely be more difficult and costly. Global warming normally a highly scientific issue has in the last many decades become a political issue. In early seventies, trillions of dollars were being spent on research to avert an impending ice age. Geologically, brief ice ages coupled with world wide regressions of seas followed prolonged phases of global warming and coincident sea level rise. The man survived the last ice age spanning briefest of all major ice ages i.e. around 2.5 million years. The first two major ice ages were Snow Ball events making entire planet covered in a blanket of snow several kilometers thick for nearly a hundred million years each time. Even during last 18000 years of scientific global warming, we have had three mini ice ages, it has been scientifically proven and accepted by global geological communities. Mini global warming and global cooling has alternated even during this period succeeding the last major ice age in Earth's History. These fascinating scientific facts of Earth history have been totally knocked out by the massive hype of manmade global warming now cleverly called climate change. Of course, land and ocean temperature is only one way to measure the effects of climate change. A warming world also has the potential to change rainfall and snow patterns, increase droughts and severe storms, reduce lake ice cover, melt glaciers, increase sea levels, and change plant and animal behavior. The Earth's sea level has risen by 21 cm (8 inches) since 1880. The rate of rise is accelerating and is now at a pace that has not been seen for at least 5000 years. Global warming has caused this by affecting the oceans in two ways, warmer average temperatures cause ocean waters to expand (thermal expansion) and the accelerated melting of ice and glaciers increase the amount of water in the oceans. Of course, land and ocean temperature is only one way to measure the effects of climate change. A warming world also has the potential to change rainfall and snow patterns, increase droughts and severe storms, reduce lake ice cover, melt glaciers, increase sea levels, and change plant and animal behavior.

ISBN-13: 978-1533168078
ISBN-10: 1533168075

Chemistry of Industrial Globalization, Environmental Pollution and its Chem-Biological Significance

February 2016

Industrial Globalization and Environmental Awareness

Proceedings of the NATIONAL SEMINAR held by

Department of Chemistry, Government PG College, Ambala Cantt - Haryana - India

Design and Synthesis of Some New Curcumin Analogs and Their Pyrazole Derivatives of Potential Biological Interest

Shilpy Aggarwal[1] and Deepika Saini[2]

[1]RKSD (PG) College, Kaithal

[2]Institte of Pharmaceutical Sciences, Kurukshetra University, Kurukshetra

E-mail- shilpysingla@rediffmail.com

ABSTRACT

Curcumin is the principal curcuminoid of the popular Indian spice turmeric (Curcuma longa), which is a member of the ginger family (Zingiberaceae) and pyrazole refers both to the class of simple aromatic ring organic compounds of the heterocyclic series characterized by a 5-membered ring structure composed of three carbon atoms and two nitrogen atoms in adjacent positions and our present study corresponds to synthesis of pyrazole derivatives of synthesized curcumin analogs. The synthesis of substituted 8-aryloct- 7-ene- 2,4,6-triones 5a-k was carried as: Chalcone analogs (3a-k) were prepared by condensation of dehydroacetic acid (DHA, 1) with various aromatic aldehyde (2a-k) in chloroform with few drops of piperidine. 3 has undergone rearrangement to 2-methyl- 6-styryl- 4-pyrones (4) under acidic conditions. When pyrones 4 were treated with $Ba(OH)_2$, converted to triones 5 in aqueous ethanol. All of these compounds 3-5 were identified by their spectral data, IR and NMR. An equimolar mixture of trione 5 and 2-hydrazino-4- methylquinoline (6) was refluxed in ethanol to give pyrazole derivative (7). In docking studies, Compound 7f showed very good affinity towards the receptor through binding to the residues HFP 2001 & HOH 1149 with highest dock score of -8.82. 7e was having highest zone of inhibition of 23mm in concentration of 100μg/ml against gram –ve bacteria. Compounds 5a, 5d, 5g & 5k exhibited significant antimalarial activity with the percent inhibition of 61.59, 61.9, 59.71 & 58.12 respectively.

ISBN-13: 978-1533168078
ISBN-10: 1533168075

Chemistry of Industrial Globalization, Environmental Pollution and its Chem-Biological Significance February 2016

Industrial Globalization and Environmental Awareness

Proceedings of the NATIONAL SEMINAR held by

Department of Chemistry, Government PG College, Ambala Cantt - Haryana - India

Physio-chemical studies of water-quality parameters of drinking water-"A case study of Jind City, Haryana (India)"

J.P. Deshwal[1], B.R. Deshwal[2], J P Saharan[3]

[1]Department of Chemistry, K.M. Government College, Narwana, Jind (Haryana) India
[2]Department of Chemistry All India "Jat Heroes" Memorial College, Rohtak (Haryana)India
[3]Department of Chemistry, S.D. College (Lahore), Ambala Cantt. (Haryana) India.
jp_govtcollege@rediffmail.com; deshwalbr@gmail.com

ABSTRACT

People on globe are under tremendous threat due to undesired changes in the physical, chemical and biological characteristics of air, water and soil. Due to increased human population, industrialization, use of fertilizers and man-made activity water is highly polluted with different harmful contaminants. Natural water contaminates due to weathering of rocks and leaching of soils, mining processing etc. It is necessary that the quality of drinking water should be checked at regular time interval, because due to use of contaminated drinking water, human population suffers from varied of water borne diseases. The availability of good quality water is an indispensable feature for preventing diseases and improving quality of life. The present study was conducted to analyze the various parameters of under-ground water in Jind City, Haryana and to check its fitness for drinking. Water samples were collected from different localities in cleaned polythene bottles. These water samples were analyzed for their physicochemical characteristics. Laboratory analyses on samples were performed for pH, Colour, Odour, Hardness, Chloride, Alkalinity, Total Dissolved Solids (TDS) and others. On comparing the results against drinking water quality standards laid by World Health Organization (WHO), it was found that some of the samples were non-potable for human consumption due to high concentrations of some of the parameters determined. An attempt was made to find whether or not the quality of ground water in the areas of study suitable for human consumption. The WQI value indicates that water samples of some sampling stations are quite unfit for drinking purpose because of high value of dissolved solids and sodium. Suitable suggestions were made to improve the quality of drinking water.

Keywords: Ground water, water quality parameter, physiochemical parameter, pollution, chemometric studies,

ISBN-13: 978-1533168078
ISBN-10: 1533168075

Chemistry of Industrial Globalization, Environmental Pollution and its February 2016 Chem-Biological Significance

Industrial Globalization and Environmental Awareness
Proceedings of the NATIONAL SEMINAR held by
Department of Chemistry, Government PG College, Ambala Cantt - Haryana - India

Studies of Quality Parameters of Drinking Water-"A Case Study of Rural Areas of Kaithal (Haryana)"

Neha[1], B.R. Deshwal[2], S P Sharma[3], P. Singh[4] & J P Deshwal[5]

1 Department of Chemistry, Rajiv Gandhi College(Uchana) Kaithal (Haryana) India
2 Department of Chemistry All India "Jat Heroes" Memorial College, Rohtak (Haryana)India
4 Department of Physics, S.D. College (Lahore) Ambala Cantt. (Haryana) India
3 Department of Chemistry, Baba Mast Nath University-Rohtak (Haryana)
5 Department of Chemistry, K M Govt College-Narwana (Haryana) India
Email: nehamalik4744@yahoo.com

ABSTRACT

This paper presents a study on drinking water quality in Rural areas of Kaithal City, which involved analyses of chemical parameters of drinking water samples from different drinking water sources. The drinking water sources examined included tap water, river water and well water (deep and shallow wells). Water quality studied includes pH, chloride, nitrate and total hardness levels. The concentrations of total hardness in mg CaCO3/L and chloride were obtained by titration method while the nitrate concentration levels were determined by spectrophotometer. Tap water was found to be of high quality than other sources in terms of chemical characteristics. The study revealed that the chemical parameters of water sources did not meet the permissible World Health Organization (WHO) and Indian Bureau of Standards (IBS) levels. Examining exceedence above the WHO standards, it was revealed that most of the samples contained chloride levels above allowable WHO limits. It was recommended that drinking water sources for domestic use should be protected from pollution sources. The WQI value indicates that water samples of some sampling stations are quite unfit for drinking purpose because of high value of dissolved solids and sodium. Suitable suggestions were made to improve the quality of drinking water.

ISBN-13: 978-1533168078
ISBN-10: 1533168075

Chemistry of Industrial Globalization, Environmental Pollution and its Chem-Biological Significance February 2016

Industrial Globalization and Environmental Awareness
Proceedings of the NATIONAL SEMINAR held by
Department of Chemistry, Government PG College, Ambala Cantt - Haryana - India

SPECTROPHOTOMETRIC DETERMINATION OF W(VI) AFTER EXTRACTION OF ITS FHB COMPLEX INTO CHLOROFORM SOLVENT

Joginder
Department of Chemistry, S D College Ambala Cantt (Haryana)
Email: joginderchemistry@gmail.com

Abstract

The present research work reported the spectrophotometric determination of FHB-W(VI) in chloroform solution. 2-(2'- furyl)-3- hydroxy-4- oxo-4H- 1-benzopyran (FHB) been used for the first time as an analytical reagent for the spectrophotometric determination of tungsten. The yellow colored W(VI) – FHB complex has λ max at 415m, is extractable into chloroform from 0.06– 0.28 M $HClO_4$ medium containing 0.6 – 1.8ml of 0.02 % FHB (in acetone). Tungsten (VI) in presence of several cations, anions and complexing agents forms a yellow 1:2 complex with FHB which is stable for about 18 h in organic extract. The molar absorptivity and Sandell's sensitivity of the complex at 415nm are calculated to be 5.607×10^4 dm^3 mol^{-1} cm^{-1} and $0.00163\mu g$ W cm^{-2}, respectively. Beer's law is obeyed over the concentration range $0 – 3.1\mu g$ W VI ml^{-1}. However, the optimum range of determination of tungsten from Ringbom's curve at 415 nm is $0.47 – 2.88$ μg ml^{-1} . For $2\mu g$ W cm^{-3} the standard deviation is 0.000734 and relative standard deviation of $\pm$ 0.120% absorbance unit. The method is simple, selective, rapid and precise and has been applied to the determination of tungsten in synthetic and standard samples.

Key words: Spectrophotometric, Complex, Tungston

Government P.G. College, Ambala Cantt.

(AFFILIATED TO KURUKSHETRA UNIVERSITY, KURUKSHETRA AND NAAC ACCREDITED 'A' GRADE INSTITUTION)

NATIONAL SEMINAR on 11th February, 2016

Sponsored by: Directorate of Higher Education, Haryana

Programme Schedule

Registration and Refreshments	: 09.00 A.M.
Inaugural Session	: 10.00 A.M.

       Dignitaries on Dias
       Sh. O.P. Singh, IPS, Commissioner of Police, Ambala-Panchkula
       Prof. Ravi Shankar, Head, Dept. of Chemistry, IIT, New Delhi
       Dr Inderjeet Singh Sandhu, Professor and Dean, Chitkara University, Punjab
       Dr Kamlesh, Principal
       Dr Avtar Singh Rahi, Seminar Convener
    Lighting of Lamps before Maa Sarswati
    Floral Welcome of Guests
    Introduction of Guests by Seminar Convener
    Formal welcome of Guests by Principal

Keynote Address	Prof. Ravi Shankar	10.15 A.M.
Lecture	Prof. Inderjeet Singh Sandhu	11.00 A.M.
Address by Chief Guest		11.40 A.M.
Presenting a Token of Love to the Guests		12.10 P.M.
Vote of Thanks	Dr Rohini Singh	12.12 P.M.

Tea Break	: 12.18 P.M.
Technical Sessions (Paper Presentation by Delegates - Oral and Poster)	: 12.30 P.M.
Lunch	: 02.15 P.M.
Valedictory Session	: 03.00 P.M.

    Dignitaries on Dias
       Prof. N.S. Atri, Dept. of Botany, Punjabi University, Patiala
       Dr Subodh Kumar, Senior-most Faculty
       Dr Avtar Singh Rahi, Seminar Convener
    Floral Welcome of Guest
    Introduction of Guest by Seminar Convener
    Formal welcome of Guest by Chairperson

Valedictory Address	Prof. N.S. Atri	03.10 P.M.
Seminar Success Remarks	Dr Avtar Singh Rahi	04.00 P.M.
Presenting a Token of Love to the Guests		04.05 P.M.
Vote of Thanks	Sh Satish Garg	04.07 P.M.

Distribution of Certificates	: 04.10 P.M.
Tea	: 04.30 P.M.

www.ingramcontent.com/pod-product-compliance
Lightning Source LLC
Chambersburg PA
CBHW080618190526
45169CB00009B/3230